Methods in Molecular Biology

Series Editor
John M. Walker
School of Life and Medical Sciences
University of Hertfordshire
Hatfield, Hertfordshire, AL10 9AB, UK

For further volumes:
http://www.springer.com/series/7651

Exosomes and Microvesicles

Methods and Protocols

Edited by

Andrew F. Hill

Department of Biochemistry and Genetics, La Trobe Institute for Molecular Science, La Trobe University, Bundoora, VIC, Australia

Editor
Andrew F. Hill
Department of Biochemistry and Genetics
La Trobe Institute for Molecular Science
La Trobe University
Bundoora, VIC, Australia

ISSN 1064-3745 ISSN 1940-6029 (electronic)
Methods in Molecular Biology
ISBN 978-1-4939-8284-4 ISBN 978-1-4939-6728-5 (eBook)
DOI 10.1007/978-1-4939-6728-5

Softcover reprint of the hardcover 1st edition 2016

Printed on acid-free paper

This Humana Press imprint is published by Springer Nature
The registered company is Springer Science+Business Media LLC
The registered company address is: 233 Spring Street, New York, NY 10013, U.S.A.

Preface

Exosomes and Microvesicles: Methods and Protocols brings together a collection of methods for studying extracellular vesicles (EV). There has been significant growth in the field of EV research over the last decade as we understand more about the role of exosomes, microvesicles, and other EVs in many facets of cellular biology. This has been brought about with the emerging role of EVs in cell-cell communication and their potential as sources of disease biomarkers and a delivery agent for therapeutics.

The protocols in this volume of *Methods in Molecular Biology* cover methods for the analysis of EVs which can be applied to those isolated from a wide variety of sources. This includes the use of electron microscopy, tunable resistance pulse sensing, and nanoparticle tracking analysis. Furthermore, analysis of EV cargoes containing proteins and genomic material is covered in detailed chapters that contain methods for proteomic and genomic analysis using a number of different approaches. Also presented are approaches for isolating EVs from different sources such as platelets and neuronal cells and tissues. Combined these provide a comprehensive discussion of relevant methodologies for researching EVs. As with other volumes in the *Methods in Molecular Biology* series, the notes sections at the end of each methods chapter give invaluable insight into the methods and provide information which can help with troubleshooting and further experimental optimization.

I would like to thank the chapter authors for their contributions to this volume and the editorial assistance of John Walker (Series Editor) in putting this volume together.

Melbourne, Australia ***Andrew F. Hill***

Contents

Contributors

MARIA AATONEN • *Division of Biochemistry and Biotechnology, Faculty of Biological and Environmental Sciences, University of Helsinki, Helsinki, Finland*
SAMIR E.L. ANDALOUSSI • *Department of Physiology, Anatomy and Genetics, University of Oxford, Oxford, UK; Department of Laboratory Medicine, Karolinska Institutet, Huddinge, Sweden*
ZAIRA E. ARELLANO-ANAYA • *IHAP, Université de Toulouse, INRA, ENVT, Toulouse, France*
RIKKE BÆK • *Department of Clinical Immunology, Aalborg University Hospital, Aalborg, Denmark*
ARJAN D. BARENDRECHT • *Department of Clinical Chemistry and Haematology, University Medical Center Utrecht, Utrecht, The Netherlands*
SHAYNE A. BELLINGHAM • *Department of Biochemistry and Molecular Biology, The University of Melbourne, Melbourne, VIC, Australia; Bio21 Molecular Science and Biotechnology Institute, The University of Melbourne, Melbourne, VIC, Australia*
EMILIEN BERNARD • *Hôpital Neurologique Pierre Wertheimer, Bron-Lyon, France*
ANITA BÖING • *Laboratory of Experimental Clinical Chemistry, Academic Medical Centre of the University of Amsterdam, Amsterdam, The Netherlands*
ALAIN R. BRISSON • *Molecular Imaging and NanoBioTechnology, UMR-5248-CBMN, CNRS-University of Bordeaux-IPB, Pessac, France*
MARIKE L.D. BROEKMAN • *Department of Neurosurgery, University Medical Center Utrecht, Utrecht, The Netherlands; Brain Center Rudolf Magnus, University Medical Center Utrecht, Utrecht, The Netherlands*
LESLEY CHENG • *Department of Biochemistry and Molecular Biology, The University of Melbourne, Melbourne, VIC, Australia; Department of Biochemistry and Genetics, La Trobe Institute for Molecular Science, La Trobe University, Melbourne, VIC, Australia*
S.M. VAN DOMMELEN • *Department of Clinical Chemistry and Haematology, University Medical Center Utrecht, Utrecht, The Netherlands*
O. ENIOLA-ADEFESO • *Department of Chemical Engineering, University of Michigan, Ann Arbor, MI, USA*
M. FISH • *Department of Chemical Engineering, University of Michigan, Ann Arbor, MI, USA*
LAURE GALLAY • *CNRS UMR5239, LBMC, Ecole Normale Supérieure de Lyon, Lyon, France; Institut NeuroMyoGène (INMG), CNRS UMR5310 – INSERM U1217, Université de Lyon – Université Claude Bernard, Lyon, France*
LAHIRU GANGODA • *Department of Biochemistry and Genetics, La Trobe Institute for Molecular Science, La Trobe University, Melbourne, VIC, Australia*
S.A. GAUTHIER • *Department of Psychiatry, New York University Langone Medical Center, Orangeburg, NY, USA; Department of Biochemistry and Molecular Pharmacology, New York University Langone Medical Center, Orangeburg, NY, USA; Division of Analytical Psychopharmacology, Center for Dementia Research, Nathan S. Kline Institute for Psychiatric Research, Orangeburg, NY, USA; Division of Neurochemistry, Nathan S. Kline Institute for Psychiatric Research, Orangeburg, NY, USA*

YONG SONG GHO • *Department of Life Sciences, Pohang University of Science and Technology, Pohang, Republic of Korea*

BERND GIEBEL • *Institute for Transfusion Medicine, University Hospital Essen, University Duisburg-Essen, Essen, Germany*

CÉLINE GOUNOU • *Molecular Imaging and NanoBioTechnology, UMR-5248-CBMN, CNRS-University of Bordeaux-IPB, Pessac, France*

DAVID W. GREENING • *Department of Biochemistry and Genetics, La Trobe Institute for Molecular Science, La Trobe University, Bundoora, VIC, Australia*

CLEMENS HELMBRECHT • *Particle Metrix GmbH, Meerbusch, Germany*

FIONA HEMMING • *Equipe 2, Neurodégénérescence et Plasticité, INSERM, U836, Grenoble, France; Grenoble Institute of Neuroscience, Université Joseph Fourier, Grenoble, France*

ANDREW F. HILL • *Department of Biochemistry and Genetics, La Trobe Institute for Molecular Science, La Trobe University, Bundoora, VIC, Australia*

CHARLOTTE JAVALET • *Equipe 2, Neurodégénérescence et Plasticité, INSERM, U836, Grenoble, France; Grenoble Institute of Neuroscience, Université Joseph Fourier, Grenoble, France*

JENNIFER C. JONES • *National Cancer Institute, National Institutes of Health, Bethesda, MD, USA; Molecular Immunogenetics and Vaccine Research Section Vaccine Branch, CCR, Bethesda, MD, USA*

MALENE MØLLER JØRGENSEN • *Department of Clinical Immunology, Aalborg University Hospital, Aalborg, Denmark*

SHIVAKUMAR KEERTHIKUMAR • *Department of Biochemistry and Genetics, La Trobe Institute for Molecular Science, La Trobe University, Melbourne, VIC, Australia*

A. KUMAR • *Department of Psychiatry, New York University Langone Medical Center, Orangeburg, NY, USA; Department of Biochemistry and Molecular Pharmacology, New York University Langone Medical Center, Orangeburg, NY, USA; Division of Analytical Psychopharmacology, Center for Dementia Research, Nathan S. Kline Institute for Psychiatric Research, Orangeburg, NY, USA; Division of Neurochemistry, Nathan S. Kline Institute for Psychiatric Research, Orangeburg, NY, USA*

KARINE LAULAGNIER • *Equipe 2, Neurodégénérescence et Plasticité, INSERM, U836, Grenoble, France; Grenoble Institute of Neuroscience, Université Joseph Fourier, Grenoble, France*

PASCAL LEBLANC • *CNRS UMR5239, LBMC, Ecole Normale Supérieure de Lyon, Lyon, France; Institut NeuroMyoGène (INMG), CNRS UMR5310 – INSERM U1217, Université de Lyon – Université Claude Bernard, Lyon, France*

YI LEE • *Department of Physiology, Anatomy and Genetics, University of Oxford, Oxford, UK*

SYLVAIN LEHMANN • *IRB, Hôpital St Eloi, Montpellier, France*

E. LEVY • *Department of Psychiatry, New York University Langone Medical Center, Orangeburg, NY, USA; Department of Biochemistry and Molecular Pharmacology, New York University Langone Medical Center, Orangeburg, NY, USA; Division of Analytical Psychopharmacology, Center for Dementia Research, Nathan S. Kline Institute for Psychiatric Research, Orangeburg, NY, USA; Division of Neurochemistry, Nathan S. Kline Institute for Psychiatric Research, Orangeburg, NY, USA*

ROMAIN LINARES • *Molecular Imaging and NanoBioTechnology, UMR-5248-CBMN, CNRS-University of Bordeaux-IPB, Pessac, France*

SYBREN L.N. MAAS • *Department of Neurosurgery, University Medical Center Utrecht, Utrecht, The Netherlands; Brain Center Rudolf Magnus, University Medical Center Utrecht, Utrecht, The Netherlands*

IMRE MÄGER • *Department of Physiology, Anatomy and Genetics, University of Oxford, Oxford, UK; Institute of Technology, University of Tartu, Tartu, Estonia*
SURESH MATHIVANAN • *Department of Biochemistry and Genetics, La Trobe Institute for Molecular Science, La Trobe University, Melbourne, VIC, Australia*
KYM MCNICHOLAS • *Flinders Centre for Innovation in Cancer, School of Medicine, Flinders University, South Australia, Australia*
MICHAEL Z. MICHAEL • *Flinders Centre for Innovation in Cancer, School of Medicine, Flinders University, South Australia, Australia; Department of Gastroenterology and Hepatology, Flinders Medical Centre, South Australia, Australia*
AIZEA MORALES-KASTRESANA, • *National Cancer Institute, National Institutes of Health, Bethesda, MD, USA*
RIENK NIEUWLAND • *Laboratory of Experimental Clinical Chemistry, Academic Medical Centre of the University of Amsterdam, Amsterdam, The Netherlands*
JOEL Z. NORDIN • *Department of Laboratory Medicine, Karolinska Institutet, Huddinge, Sweden*
R. PEREZ-GONZALEZ • *Department of Psychiatry, New York University Langone Medical Center, Orangeburg, NY, USA; Department of Biochemistry and Molecular Pharmacology, New York University Langone Medical Center, Orangeburg, NY, USA; Division of Analytical Psychopharmacology, Center for Dementia Research, Nathan S. Kline Institute for Psychiatric Research, Orangeburg, NY, USA; Division of Neurochemistry, Nathan S. Kline Institute for Psychiatric Research, Orangeburg, NY, USA*
MONIQUE PROVANSAL • *IRB, Hôpital St Eloi, Montpellier, France*
GRAÇA RAPOSO • *CNRS UMR144, Institut Curie, Paris, France*
RÉMY SADOUL • *Equipe 2, Neurodégénérescence et Plasticité, INSERM, U836, Grenoble, France; Grenoble Institute of Neuroscience, Université Joseph Fourier, Grenoble, France*
MARIKO SAITO • *Division of Neurochemistry, Nathan S. Kline Institute for Psychiatric Research, Orangeburg, NY, USA; Department of Psychiatry, New York University Langone Medical Center, New York, NY, USA*
MITSUO SAITO • *Division of Analytical Pshycopharmacology, Nathan S. Kline Institute for Psychiatric Research, Orangeburg, NY, USA*
LAURENT SCHAEFFER • *CNRS UMR5239, LBMC, Ecole Normale Supérieure de Lyon, Lyon, France; Institut NeuroMyoGène (INMG), CNRS UMR5310 – INSERM U1217, Université de Lyon – Université Claude Bernard, Lyon, France*
R.M. SCHIFFELERS • *Department of Clinical Chemistry and Haematology, University Medical Center Utrecht, Utrecht, The Netherlands*
MITCH SHAMBROOK • *Department of Biochemistry and Genetics, La Trobe Institute for Molecular Science, La Trobe University, Melbourne, VIC, Australia*
PIA SILJANDER • *Division of Biochemistry and Biotechnology, Faculty of Biological and Environmental Sciences, University of Helsinki, Helsinki, Finland*
RICHARD J. SIMPSON • *Department of Biochemistry and Genetics, La Trobe Institute for Molecular Science, La Trobe University, Melbourne, VIC, Australia*
SISAREUTH TAN • *Molecular Imaging and NanoBioTechnology, UMR-5248-CBMN, CNRS-University of Bordeaux-IPB, Pessac, France*
PIETER VADER • *Department of Physiology, Anatomy and Genetics, University of Oxford, Oxford, UK; Department of Clinical Chemistry and Haematology, UMC Utrecht, Utrecht, The Netherlands*

Sami Valkonen • *Laboratory of Experimental Clinical Chemistry, Academic Medical Centre of the University of Amsterdam, Amsterdam, The Netherlands*
Didier Vilette • *IHAP, Université de Toulouse, INRA, ENVT, Toulouse, France*
Jeroen de Vrij • *Erasmus Medical Center, Rotterdam, The Netherlands; Department of Neurosurgery, University Medical Center Utrecht, Utrecht, The Netherlands; Brain Center Rudolf Magnus, University Medical Center Utrecht, Utrecht, The Netherlands*
Matthew J.A. Wood • *Department of Physiology, Anatomy and Genetics, University of Oxford, Oxford, UK*
Rong Xu • *Department of Biochemistry and Genetics, La Trobe Institute for Molecular Science, La Trobe University, Melbourne, VIC, Australia*
Yuana Yuana • *Laboratory of Experimental Clinical Chemistry, Academic Medical Centre of the University of Amsterdam, Amsterdam, The Netherlands*

Chapter 1

Methods to Analyze EVs

Bernd Giebel and Clemens Helmbrecht

Abstract

Research in the field of extracellular vesicles (EVs) is challenged by the small size of the nano-sized particles. Apart from the use of transmission and scanning electron microscopy, established technical platforms to visualize, quantify, and characterize nano-sized EVs were lacking. Recently, methodologies to characterize nano-sized EVs have been developed. This chapter aims to summarize physical principles of novel and conventional technologies to be used in the EV field and to discuss advantages and limitations.

Key words Nanoparticle tracking analysis, Electron microscopy, Dynamic light scattering, Flow cytometry, Extracellular vesicles, Resistive pulse sensing

1 Introduction

Eukaryotic and prokaryotic cells release a variety of nano- and micron-sized membrane-containing vesicles into their extracellular environment, which are collectively referred to as extracellular vesicles (EVs). EVs can be harvested from cell culture supernatants and from all body fluids including plasma, saliva, urine, milk, and cerebrospinal fluid [1]. Depending on their origin, different EV subtypes can be distinguished. Together with apoptotic bodies (1000–5000 nm), exosomes (70–160 nm) and microvesicles (100–1000 nm) provide the most prominent groups of EVs. Exosomes are defined as derivatives of the endosomal system and correspond to the intraluminal vesicles of multivesicular bodies (MVBs), which, upon fusion of the MVB with the plasma membrane, are released into the extracellular environment [2–4]. In contrast, microvesicles are directly pinched off the plasma membrane [3]. Even though the release of exosomes was initially reported in 1983 by detailed structural analysis using transmission electron microscopy [5], research on nano-sized EVs did not gain significant prominence until the discovery that small EVs transport small RNAs, including micro RNAs [6, 7]. Since then, the interest

Andrew F. Hill (ed.), *Exosomes and Microvesicles: Methods and Protocols,* Methods in Molecular Biology, vol. 1545, DOI 10.1007/978-1-4939-6728-5_1, © Springer Science+Business Media LLC 2017

in EVs as mediators for intercellular signaling, biomarkers for diseases, drug delivery vehicles, or therapeutical agents has dramatically increased [8, 9].

The research in the EV field is challenged by the small size of the nano-sized EVs. Apart from transmission and scanning electron microscopy, established technical platforms to visualize, quantify and characterize nano-sized EVs were lacking. In 2011 the nanoparticle tracking analysis (NTA) was initially described as a useful technology to characterize nano-sized EVs [10, 11]. NTA has emerged as one of the most prominent, state-of-the-art technologies in the EV field. In addition, other methods adopted from the field of nanotechnology are available, which have been or might be used for the characterization of EVs. This chapter aims to summarize physical principles of novel and conventional technologies to be used in the EV field and to discuss advantages and limitations, which are summarized in Table 1.

Table 1
Current methods for EV analysis

Technique	Particle size	Time for measurement	Limitations	Advantages
Cryo-TEM	<1 nm … mm	>1 h	Sample preparation, only small amount of sample is analyzed	Morphology
DLS, homodyne	1 nm … 6 μm	1–2 min	Polydisperse samples challenging, presence of large particles biases results	Wide size range
DLS, heterodyne	0.5 nm … 6 μm	1–2 min	Analog homodyne DLS, but not as dominant	Wide ranges of size and concentration
NTA	20 nm … 1 μm	5–10 min	Dilution necessary for high concentration, non-standardized method	Visualization, resolution (even polydisperse samples), low concentrations
FCM	300–500 nm … 10 μm	1 min	Working range, pore blocking, calibration	Fluorescence, biochemical information
AFM	10–1000 nm	>1 h	Analog cryo-TEM	Morphology
RPS	50 nm to 10 μm, dependent on pore size	30 min	Working range, pore blocking, calibration	High resolution, compatible with buffers
AF^4	ca. 5 nm to 20 μm	1 h	Sample dilution, interaction of sample with membrane	Fractionation

TEM transmission electron microscopy, *DLS* dynamic light scattering, *NTA* nanoparticle tacking analysis, *FCM* flow cytometry, *AFM* atomic force microscopy, *RPS* resistive pulse sensing, *AF4* flow field flow fractionation

2 EV Analysis

There are a number of optical and nonoptical methods to analyze the size, quality, and concentration of nanoparticles. The maximal resolution (d_A) of classical light microscopy depends on the wavelength of light (λ) and the numerical aperture (NA) of the lenses. It can be calculated according to the formula:

$$d_A = \frac{\lambda}{2 \cdot \mathrm{NA}}. \tag{1}$$

High-quality lenses (e.g., oil immersion objectives) rarely reach apertures of more than 1.4. Accordingly, at a supposed wavelength of 550 nm, conventional light microscopes have difficulty resolving structures less than 200 nm in size. To detect smaller structures, electron microscopic techniques are required. Thus, EVs are conventionally analyzed by electron microscopy, usually via transmission electron microscopy (TEM) and in some cases by scanning electron microscopy (SEM) [10, 12]. New fluorescence based super-resolution microscopic techniques such as STED (stimulated emission depletion) or PALM (photoactivated localization microscopy) and atomic/scanning force microscopy definitively allow for higher resolutions and certainly will provide important information on the nature of EVs in the near future [13–16].

2.1 Electron Microscopy

For the preparation of EV samples for electron microscopy different methods can be used. Heavy metals, such as osmium tetroxide and uranyl acetate, increase the contrast of the analyzed samples. However, like aldehyde-based fixation methods, heavy metal treatment regularly results in the dehydration of the samples, resulting in EV shrinkage and deformation. Accordingly, EVs frequently adopt cup-shaped morphologies, which were initially considered as a characteristic feature of exosomes [12]. Upon using cryoelectron microscopic technologies lacking chemical fixation and staining procedures, native EV sizes and shapes can almost be conserved. Here, freshly prepared EVs are transferred to grids and immediately are cryofixed in liquid nitrogen. As a result of the procedure, water is placed in a glass-like state without forming destructive ice crystals, thus, leaving the EV structure largely intact [17, 18]. Although the electron microscopic analyses provide important information on the EV morphology, this technology does not allow EV quantification; among others EVs do not quantitatively adhere to the grids.

2.2 Physical Background on Light Scattering

Methods based on the analysis of scattered light are eminently suitable for the contact-free analysis of delicate samples such as bio-nanoparticles—EVs. In nearly every analysis device, ranging from dynamic light scattering (DLS) to fluorescent cell sorting, light scattering is utilized to gain information about the samples in a fast and efficient way. Before describing current techniques, a brief physical background about the light scattering features of small particles should be given.

2.2.1 Tyndall Effect

When small particles (ranging in size from approximately 100 nm to several μm), such as in diluted milk or fog, are illuminated by a directed beam of light from a laser pointer, the light beam becomes visible as the particles scatter the incident light. Single, larger particles can even be recognized by eye, like dust in the sunlight. In the middle of the nineteenth century, Tyndall (1820–1893) observed this phenomenon and used it for the detection of small particles in liquids. Although he probably was not the first who discovered this phenomenon, the effect has been termed the "Tyndall effect." The scattered light contains information, allowing detection and analysis of the particles which is employed in common and new techniques based on light scattering.

2.2.2 Elastic and Inelastic Light Scattering

Bearing in mind the principle of energy conservation, energy cannot be created nor destroyed but only changed from one form into another. As an example, energy (the momentum) can be transferred from one billiard ball to another. In the elastic case, the billiard balls deform during collision (although this cannot be seen by eye), kinetic energy is transferred and the billiard balls return to the original form. In contrast, if one of the balls would be made of modeling mass, parts of the transferred energy lead to inelastic deformation of the modeling mass ball and only parts of the momentum are transferred as kinetic energy.

This example reflects the underlying principle in light scattering. Light is an electromagnetic wave with wavelengths visible to the human eye ranging from 380 to 780 nm. The electromagnetic wave consists of a large number of small discrete energy packages, the photons. The energy of a photon, transferring the energy (E), can be calculated via the expression

$$E = \frac{h\,c_0}{\lambda} \tag{2}$$

(h Planck's constant, $h = 6.626 \times 10^{-34}$ Js; c_0: speed of light in vacuum, $c_0 = 2.998 \times 10^{8}$ ms^{-1}).

Upon illumination of a given particle, the light wave interacts with the particle; more precisely, the photons of the light wave transfer their energy to the particle's electrons. As a consequence the electrons oscillate and finally energy can be released in form of scatter light in all directions uniformly. In **elastic scattering**, the energy transferred by the photons is identical to the energy of the scatter light; consequently, the incident and scatter light have identical wavelengths (Fig. 1).

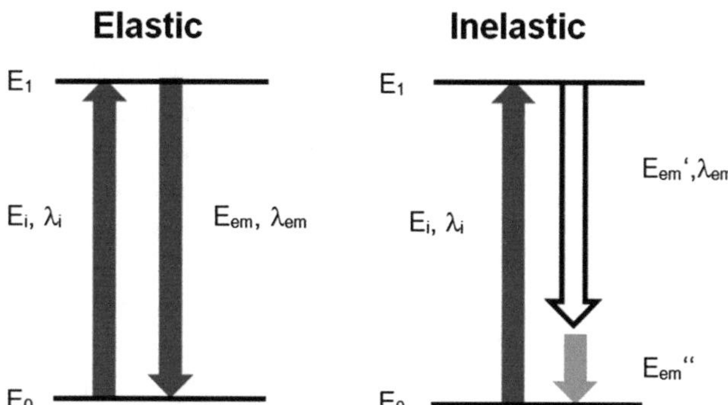

Fig. 1 Incident light wave with E_i and λ_i shifts the electrons of the particle from a ground state E_0 to a virtual level E_1. Elastic scattering: from the virtual level E_1, electrons return to the ground state, no energy is transformed (e.g., Rayleigh scattering). Inelastic scattering: electrons do not return to the ground state E_0. Parts of the incident energy (E_{em}'') are transformed into other energy forms (e.g., Stokes scattering)

Examples for elastic light scattering are **Rayleigh scattering** and **Mie scattering**, which will be explained below. If the energy of the incident and scatter light differ from each other, the process is termed **inelastic light scattering**. Light scattered in a Raman process is an example of **inelastic light scattering**. In such a process a part of the energy of the incident light is transformed into another form of energy, e.g., heat or vibrational energy. In Stokes Raman scattering, the wavelength of the scattered light is longer than the incident light. In anti-Stokes Raman scattering the wavelength of the scattered light is shorter than the wavelength of the incident light. The additional energy derives from vibrational energy of the molecules of the particle, e.g., when the molecules are in excited state.

Of note, compared to elastic scattering, **Raman scattering** is very weak and requires well thought-out arrangements for detection [19, 20].

2.2.3 The Influence of Particle Size

The characteristic of how particles scatter light is mainly related to their size. Within the scope of this chapter, we focus on Rayleigh and Mie scattering.

2.2.4 Rayleigh Scattering

Rayleigh scattering describes the elastic scattering of electromagnetic waves on particles with sizes rather small compared to the incident wavelength $r < 0.2\ \lambda$. The intensity of the scattered light is inversely related to the fourth power of the wavelength λ of the incident light. Consequently, light with shorter wavelengths is scattered with higher intensities than light with longer wavelengths. A well-known phenomenon which can be explained by Rayleigh scattering is the blue color of the sky;

molecules in the atmosphere scatter the blue parts of sunlight approximately ten times stronger than the red parts.

The scattering intensity I also depends on the index of refraction n of both, of the particle (n_1) and surrounding medium (n_2). The refractive index is defined as the ratio between the speed of light in a given material and in a vacuum. The relative refraction index $m = n_1/n_2$; n_1 and n_2 are the refractive indices of particle and surrounding media, respectively. Considering all these parameters, the intensity (I) of the Rayleigh scattering at a certain distance (R) and scattering angle (θ) [21] is given by:

$$I = I_0 \times \frac{1 + \cos^2\theta}{2R^2} \left(\frac{2\pi}{\lambda}\right)^4 \left(\frac{m^2 - 1}{m^2 + 2}\right)^2 \left(\frac{d}{2}\right)^6 \quad (3)$$

Of note, the intensity of Rayleigh scattering is proportional to the sixth power of the size of small particles, which restricts the size detection limit of many scatter based methods. In contrast, the irradiation intensity (I_0) is only linearly linked to the intensity of Rayleigh scattering. A large difference in the refractive index of the surrounding medium and the illuminated particles (e.g., water $n_2 = 1.333$) increased the intensity of the scattered light.

2.2.5 Mie Scattering

Particles with similar or larger sizes than the wavelength of the incident light cause *Mie scattering*. The formula to calculate the intensity of Mie scattering at a given angle and distance of larger particles is much more complex and is neglected here. Particles with an approximate size of the wavelength of the incident light can be considered as an aggregation of material, whose oscillating electrons influence each other and may scatter the light toward a certain direction. As a consequence, the Mie scattering intensity is less dependent on the wavelength of light than Rayleigh scattering. For example, waterdrops in clouds cause wavelength-independent Mie scattering; that is the reason why clouds appear white.

For a more detailed description on light scattering, we like to refer to more specific literature [22, 23].

3 Methods Based on Light Scattering

3.1 Dynamic Light Scattering (DLS)

An advanced technology applying the scattering light for the characterization of nanoparticles is the method of dynamic light scattering (DLS), also known as photon correlation spectroscopy (PCS). Here, a distinct proportion of the sample volume—regularly a few microliters—is illuminated with a laser beam. The light scattered from the particles within the illuminated part of the probe is recorded over time [24]. Due to their Brownian motion, the particles in the sample are constantly moving, some of them leaving and some of them entering the illuminated part of the probe. This causes fluctuations of the scattering light, which is registered by the detector. Since smaller particles move faster within the probe than larger particles, smaller particles cause higher fluctuations than larger particles. By the combination of mathematical models of the Brownian motion and the light scattering theory differential particle sizes can be calculated within seconds [25]. While in the beginning of commercial DLS (around 1970) only narrow size distributions could be measured, the range of modern DLS instruments typically covers sizes ranging between 1 nm and 6 μm [26]. To obtain optimal results, the presence of contaminants such as dust particles, air bubbles, debris and inorganic particles, which can derive from laboratory water (e.g., silicates, phosphates, carbonates), must be circumvented. For better reproducibility, optimized sample preparation including filtration of buffers is mandatory [27].

Depending on the position of the detector, two different DLS systems are commercially available, the homodyne and the heterodyne DLS.

In a homodyne DLS setup, the laser and detector are arranged perpendicular to each other (Fig. 2). The incident light with the intensity I_0 illuminates the sample and becomes partially scattered by the particles suspended in the probe. The intensity of the scattered light (I_S) is recorded by the detector. Critical parameters in this setting are the distance the light has to pass through the sample until it reaches the detector and the concentration of the particles. If the particles are too concentrated, secondary scattering occurs diminishing the amount of scatter light that reaches the detector. Hence, appropriate dilutions have to be titrated to obtain valid data [28].

Within heterodyne DLS systems the backscattered light is analyzed (Fig. 2). The incident laser light is coupled into an optical fiber to illuminate the probe with the intensity I_0. Only light, which is scattered by the particles within the probe in an angle of 180°, can reenter the optical fiber and become transmitted with

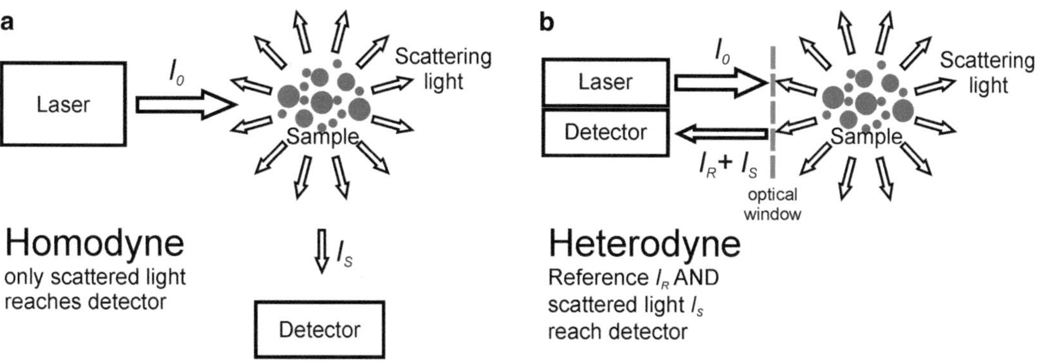

Fig. 2 Principle of homodyne and heterodyne DLS systems

an intensity I_S to the detector [29]. In addition to the average size distribution of the particles in the probe and following calibration, heterodyne DLS regularly enables to determine the particle concentration of given probes. For appropriate measurements of particle sizes, analyses of polydisperse probes require particle size differences with ratios of $d_1/d_2 > 1.8$ [30].

Regularly, a 20–50 μL sample volume is sufficient to determine the average particle size distribution on commercial DLS instruments in less than a minute. Analyses of monodisperse samples, i.e., samples only containing particles with the same size, yield reliable results. In the case of polydisperse samples such as blood plasma samples, the results may be less clear and require knowledge of the applicable mathematical model. The results are distorted by larger particles with diameters in the micrometer range, already when they are present at low concentrations [30]. Upon analyzing samples with high particle concentrations or samples containing larger agglomerates, heterodyne DLS instruments provide more flexibility than homodyne instruments, but still are limited compared to other techniques such as the *nanoparticle tracking analysis* (NTA) [31].

3.2 Nanoparticle Tracking Analysis (NTA)

In 2011 NTA was reported to provide a suitable method for EV characterization for the first time [10, 11]. Since then, NTA has emerged as one of the standard techniques for the characterization of EVs. It also allows analyses of larger particles within the micrometer range and thus has also been designated as particle tracking analysis (PTA).

Analogous to DLS, NTA records the Brownian motion of small particles. Similar to DLS, particles in the sample are visualized by the illumination with incident laser light. The scattered light of the particles is recorded with a light-sensitive CCD camera, which is arranged at a 90° angle to the irradiation plane (Fig. 3). The 90° arrangement, also known as ultramicroscopy, allows detection and tracking of the Brownian motion of 10–1000-nm-sized vesicles. Using a special algorithm the size of each individually tracked particle is calculated, thus simultaneously allowing determination of the average size distribution of particles in a given sample as well as their concentration. Even though the NTA technology is relatively new on the market, it originated almost 25 years ago [32]; the commercial implementation of this technique required the availability of fast computer systems that are able to cope with the computationally intensive video analysis in reasonable time frames.

A brief introduction of the physical principle underlying NTA is as follows: When small particles are dispersed in a liquid (the so-called continuous phase, e.g., water), the particles move randomly in all directions. This phenomenon is termed diffusion and is expressed by the diffusion coefficient (D). In more detail, the

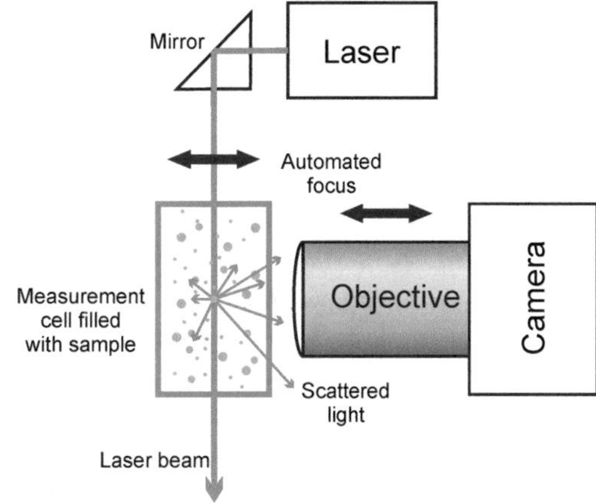

Fig. 3 Schematic setup of a nanoparticle tracking analyzer

undirected migration of given particles is caused by energy transfers from surrounding water molecules to the particle. In the absence of any concentration gradient within the dispersion and upon long-term observation, the distances small particles move in any direction should neutralize each other over time, leaving a total movement of almost zero. However, during given time intervals, diffusing particles move within certain volume elements. In NTA the time t between two observation spots is quite short (~30 ms). The distance particles have moved during the time interval are recorded and quantified as the mean square displacement (x^2). Depending on the number of dimensions (one, two or all three dimensions) the diffusion coefficient can be calculated from the mean square displacement as follows:

$$D = \frac{\langle x \rangle^2}{2t}\; D = \frac{\overline{\langle x, y \rangle}^2}{4t}\; D = \frac{\overline{\langle x, y, z \rangle}^2}{6t}.$$

Via the Stokes-Einstein relationship, the particle diameter d can be calculated as function of the diffusion coefficient D at a temperature T and a viscosity η of the liquid (k_B Boltzmann's constant) [33]:

$$D = \frac{4k_B T}{3\pi\eta d}.$$

In NTA, the particle fluctuation of a single particle is registered in two dimensions. After combining the Stokes-Einstein relationship and the two-dimensional mean square displacement, the equation can be solved for the particle diameter d with:

$$d = \frac{4k_B T}{3\pi\eta t} \cdot \frac{4t}{\overline{\langle x, y \rangle}^2} = \frac{16k_B T}{3\pi\eta\overline{\langle x, y \rangle}^2}.$$

By simultaneously tracking several particles, their diameters can be determined in parallel. Figure 4 shows a typical particle size distribution of vesicles harvested from blood plasma.

The lower limit of the working range, i.e., the smallest detectable particle size, depends on the scattered intensity of the particle (compare Eq. 3), the efficiency of the magnifying optics and the sensitivity of the camera [34]. Silver and gold nanoparticles are strong scatterers due to the comparably large refractive indices of 2–4 and can be detected down to sizes of ~10 nm. Biological nanoparticles such as EVs have refractive indices of around 1.37–1.45 resulting in a limit of detection of 30–50 nm for NTA [35].

NTA allows the direct measurement of concentration as single particles in the illuminated volume are visualized. Thus, NTA is an absolute measurement technique allowing the determination of total surface or volumes of particles in a sample (*see* Fig. 4). For the measurement of concentration, the instrument is calibrated with

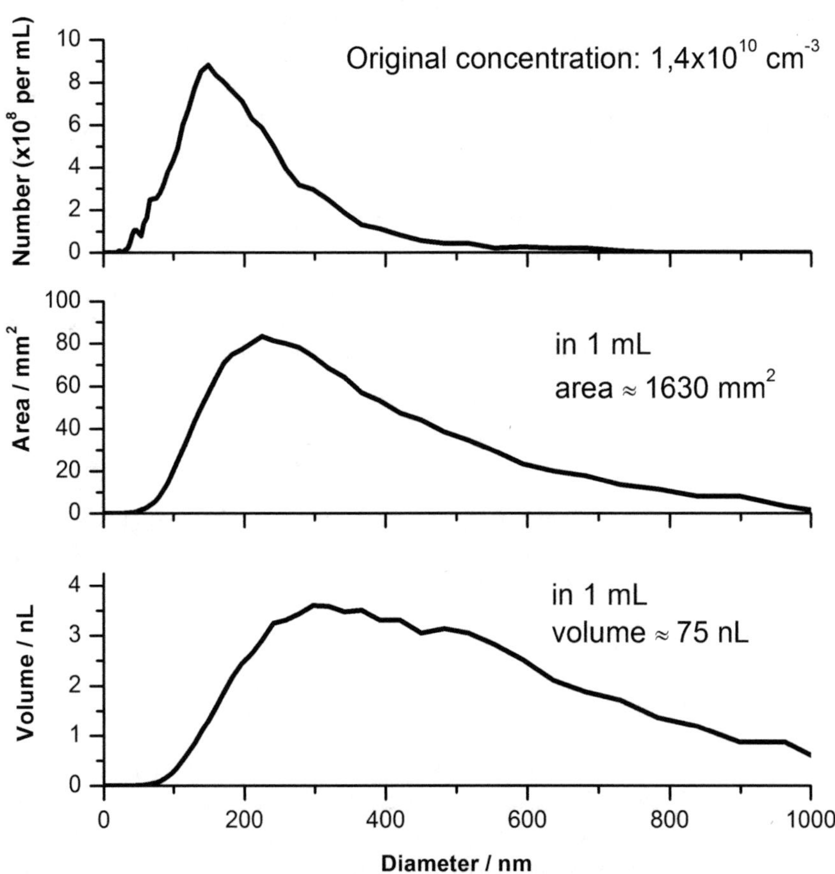

Fig. 4 Particle size distributions of vesicles in blood plasma. The particle size distributions range from <100 to 1000 nm dependent on weighing according to number, area, or volume. NTA as absolute technique allows quantification of concentration, area, and volume of vesicles present in the sample

size standards of known size and concentration. The visualization of the sample gives a unique impression on the quality of the sample, such as the presence of agglomerates. The working range of 0.5×10^6 and 1×10^{10} particles per cm^3 is very low compared to DLS, allowing NTA to analyze low concentrated samples. To record representative size distribution profiles, it is recommended to analyze a range of 1000–10,000 single particles.

While in the early stages of NTA development, the manual adjustment of microscope and laser was time-consuming, nowadays, the measurement cell is aligned within minutes. Currently, commercial NTA instruments are offered by only two companies (Malvern Instruments Ltd. and Particle Metrix GmbH). Depending on the model temperature control, conductivity and zeta potential measurement are integrated. The zeta potential reflects the surface charge of given particles, which might be related to their stability. Currently, efforts are undertaken to implement additional components, which, for example, can automatically dilute probes to optimal particle concentrations, record electrochemical parameters (e.g., the pH of the probe), and allow for the specific characterization of fluorescent-labeled EVs.

The quality of an NTA result is influenced by particle contamination. In addition to the contaminating particles, which were mentioned in the section of DLS, high concentrations of stabilizing agents (e.g., surfactants) are critical as soon as they reach their critical micellar concentration (CMC). Contaminating particles may derive from diluents (distilled water or buffer agents) or from chemicals used during preparation of samples. Regularly, chemicals are not certified for the absence of nanoparticles. Precipitates of phosphates, carbonates, or silicates as well as dust can be removed by filtration of the buffers, ideally with pore sizes below 50 nm. Degassing in ultrasonic bath is also helpful to remove air bubbles [34].

3.3 Flow Cytometry (FC)

For the characterization of EVs, it would be desirable to simultaneously analyze the presence of different molecules expressed on the surface of EVs using a high-throughput technology. At the cellular level, such analyses are regularly performed by FC. However, due to the configuration of conventional flow cytometers, the size detection limits of particles lie somewhere between 300 and 500 nm [36]. Thus, by means of conventional flow cytometry, only large EVs can be analyzed at an individual particle level. To this end, EV FC analyses have indeed already been carried out on larger EVs, particularly in the area of platelet research. In the literature corresponding EVs are usually referred to as microparticles [37–39]. Analyses of smaller EVs by flow cytometry require either a special mechanical setup, or EVs must be bound by immunological methods to carrier particles.

Magnetic carrier particles or latex beads can be coated with antibodies that recognize epitopes on EVs, e.g., anti-CD63 antibodies. If the antibody-coated beads are added to EV-containing samples, aggregates between the beads and the EVs are formed, which can be concentrated by magnetic separation or by low-speed centrifugation, respectively. For an appropriate aggregation, sufficient quantities of EVs need to be present in the sample; the beads should get saturated with EVs, otherwise aggregates with several beads might form. The aggregate formation of EVs with several beads can be reduced by vortexing or pipetting. In analogy to cells, the formed bead-EV aggregates can be labeled with different fluorescence-labeled antibodies. Due to the presence of the beads, these aggregates are big enough to be analyzed on conventional flow cytometers [40–43]. This technology offers the great advantage for a fast and comprehensive EV characterization. However, since only aggregates and not individual EVs are analyzed, this form of analysis is a bulk analysis and finally may not reveal much more information than conventional Western blots.

Irrespective of the low size resolution of conventional flow cytometers, analyses of small EVs at the single-particle level provide several challenges. As long as the particles are larger than the wavelength of light, their size corresponds to the amount of the forward-scattered light, which is measured at the forward scatter detector. If the particle sizes are around or below the wavelength of the light, the intensity of light scattered to the side increases proportionally to the forward-scattered light. Accordingly, the size of particles that are smaller than the wavelength of the incident light can better be determined upon measuring the scattered light at the side scatter detector than on the forward scatter detector. Alternatively, an extended forward scatter detector can be used, which collects the forward-scattered light and proportions of the side scattered light.

Groups that have optimized the setup of configurable flow cytometers for the measurement of nano-sized particles were already able to analyze viruses and EVs at a single-particle resolution [44–46]. Essential prerequisites for such measurements are the reduction of signal-to-noise ratio and an increase in the sensitivity of the scatter light detection. According to the formula of the Rayleigh scattering, a linear increase in sensitivity can be achieved by increasing the intensity of the laser light [44]. In addition, the signal-to-noise ratio largely depends on the processing of the sheath fluid. Regularly, commercial products are sterilized by filtration through 0.22 μm filters, which is not sufficient to remove background noise producing nanoparticles such as calcium phosphate or calcium carbonate nanoparticles. Thus, filtration through 0.05 μm filters is highly recommendable [44]. The background noise can also be reduced upon staining EVs with a strong fluorescent dye, e.g., the membrane-intercalating PKH67, and by triggering the subsequent flow cytometric measurements on the fluorescence and not as conventionally on the scattered light [46]. The disadvantage here is that aggregates of the unbound fluorochromes should be removed before stained EVs get analyzed. Even though it is time consuming, currently, density gradient centrifugation appears as the most appropriate technology to separate fluorochrome aggregates and stained EVs. Irrespectively of this, EVs can also be marked with fluorescence conjugated antibodies allowing for the specific analyses of antigens of interest [46, 47]. Since the surface of EVs is orders of magnitude smaller than that of cells, antibodies should be used being conjugated to very bright fluorochromes such as B-phycoerythrin (B- PE) or R-PE. Usage of antibodies with weaker fluorochromes can only be recommended, when corresponding epitopes are known to be expressed on the EVs very abundantly [47].

Another challenge is the concentration of the EVs to be measured. Ideally, for single particle analyses, the concentration of particles to be measured should be in the range of 5×10^5 to 5×10^6 particles per ml sample liquid. If particles are higher concentrated, swarm detection can occur, that is, the simultaneous detection of several particles at a given moment [48]. Following enrichment of EVs, the concentration regularly strongly exceeds this value; consequently, probes to be measured have to be diluted to sometimes homeopathic appearing dilutions.

3.4 Raman Microspectroscopy (RM)

Raman scattering is a form of inelastic light scattering [19]. Even though most of the incident light is scattered in an elastic manner, each molecule also specifically scatters light in an inelastic manner and thus generates individual Raman spectra of the scattered light. Raman microspectroscopy allows the recording and analysis of sample spectra and thus gives information on molecular composition of probes of interest. This technique has been used to analyze the composition of EVs and allowed discrimination of different EV subtypes from each other [49]. Especially when combined with atomic force microscopy, Raman spectroscopy might offer a very potent technology to analyze and discriminate different EV subtypes [50].

Raman microspectroscopy is a relatively high-priced and specialized technique. Setup and acquisition require a relatively large amount of time, resulting in an incompatibility with high-throughput analyses (10–100 vesicles per hour). Due to the low intensity of the Raman scattering signal (approx. <1:10,000 of elastic scattering), the measurement is influenced by artifacts demanding high grade of manual effort and expertise of the operating personnel. During measurement, EVs are exposed to a high-intensity light beam, which can induce photostress and cause adverse effects. Depending on the dose and wavelength of the incident light beam, (photo) reactions might be induced in the EVs and change them irreversibly [49].

3.5 Scattered-Light-Independent Technologies

3.5.1 Atomic Force Microscopy (AFM)

In the 1980s, considerable efforts were made to develop techniques allowing resolving solid state surfaces at atomic levels. As a result, the atomic force microscope (AFM) [51] and later the scanning tunneling microscope (STM) were developed.

AFM is based on a tip mounted on a cantilever that is moved like the pick-up of a record player in a defined distance over the surface of the material to be analyzed. The radius of the tip ideally is reduced to that of a few atoms. The torsion of the cantilever is a measure for the forces between tip and surface as function of the distance. The tip is either attracted (e.g., van der Waals forces) or repelled (e.g., electrostatic forces) from the surface resulting in characteristic force-distance curves. In the beginning, AFM has been utilized for the quantitative description of the topology of solid-state surfaces under vacuum conditions. Meanwhile immobilized particles such as vesicles can also be analyzed in buffers [36, 52, 53]. Thus, AFM became a feasible method for the characterization of EVs, especially to analyze their size and topology [54, 55]. However, as immobilization of EVs might affect their topology, results are influenced by the mode of sample preparation [56].

3.5.2 Resistive Pulse Sensing (RPS)

Resistive pulse sensing (RPS) is a technology to measure absolute sizes and the concentration of particles in suspension, whose sizes range from 100 nm to 100 μm. In principal, the system contains two cells, both equipped with an electrode. The cells are connected by membrane containing a small pore or a micro-channel, regularly with pore sizes below 1 μm (Fig. 5). To analyze the particle concentration and the average size distributions of suspensions, an electric field is applied onto the electrodes. As a consequence, charged particles migrate to the anode or cathode, respectively. In analogy to the Coulter principle, each time a particle passes through the pore, the electrical resistance of the buffer gets altered. These alterations in resistance are recorded. Since alterations in the resistance depend on the volume of the migrating particles, the particle sizes and their zeta potential can be calculated [57]. As a prerequisite for this method, the pore diameter (q) has to be much smaller than the pore thickness (l). Following calibration with particles of defined sizes, particle sizes and their zeta potentials can be calculated; they are proportional to the shapes and heights of the recorded pulses. Considering the pore of the membrane as a cylinder, the electrical resistance (R) of the pure buffer can be calculated as:

$$R = \rho \frac{l}{A}$$

ρ: specific resistance of the buffer, l: pore thickness (typically several tens of μm), A: pore area.

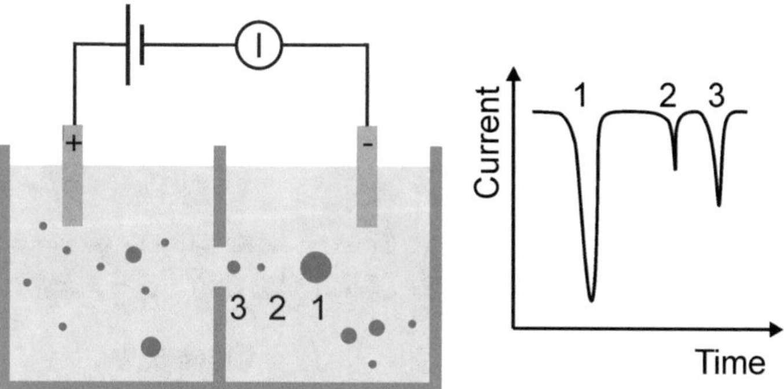

Fig. 5 Resistive pulse sensing (RPS). Left: Typical setup with two cells separated via an insulating membrane with a single pore. Right: Transient signal of current representing a (1) large-, (2) small-, and (3) medium-sized particle. Following calibration, the count rate, i.e., the number of pulses per time interval, reflects the concentration of the particle suspension to be analyzed

With $A = \pi/4 \times q^2$ the pore area is related to the pore diameter (q). In reality, each particle contains a specific electrical resistance which theoretically has to be considered. However, specific electrical resistances of given particles are high. If particles are considered as insulators, their specific electrical resistance can be neglected [58].

Provided the platform is equilibrated with particles of known concentration, the estimated count rates of given particle suspensions to be analyzed reveal their particle concentrations.

The upper end of the working range is limited by the pore size, the lower end on the sensitivity in the detection of resistance changes (typically ~0.2 q). Before usage, every membrane needs to be calibrated with size standards. Upon analyzing biological samples, pore blocking often increases the analysis time per sample of up to 1 h. RPS instruments capable of detecting particles in the lower nanometer ranges (in general <100 nm) are under development [59].

3.5.3 Field Flow Fractionation (FFF)

The family of field flow fractionation techniques (FFF) comprises instruments separating polydisperse samples in individual fractions while simultaneously determining their particle size. FFF techniques are characterized by high resolution and compatibility to flow detectors and have already been used to characterize EVs [60, 61].

The separation is based on the so-called cross-flow principle, in which two orthogonal forces act on the particle. Depending on the underlying FFF technique, the forces can be created differently, either by friction in flow field FFF (FFFF, F^4), sedimentation (sedimentation FFF, SdFFF), or an electrical field (ElFFF). FFFF and SdFFF are the most common techniques [62].

In FFFF a separation channel with an asymmetrical flow profile (asymmetrical FFFF, AF^4) is prevailingly used; it gives the most reproducible results with lowest sample loss (Fig. 6). Before

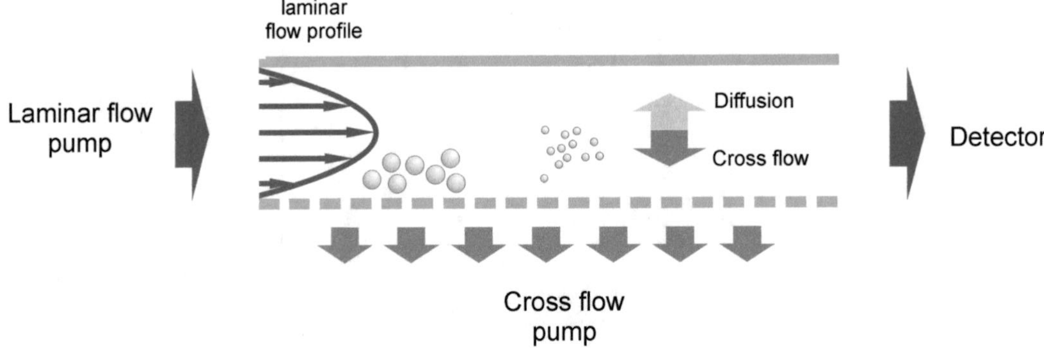

Fig. 6 Asymmetrical flow field flow fractionation (AF^4). Before separation, given samples are concentrated on the membrane. Depending on their sizes, particles diffuse from the membrane and accumulate at certain positions in which equilibriums of diffusing and cross flow forces are given. Simultaneously to their diffusion, the laminar flow transports the particles toward the detector at the end of the flow channel

fractionation starts, the sample is focused as a small band on the semipermeable membrane. During fractionation two perpendicular flows act on the sample, a laminar flow and a cross flow. The cross flow counteracts the diffusion tendency of the particles away from the membrane. Approximately 10 min after starting the measurement, the diffusion and cross flow are equilibrated and the particles have accumulated in a certain distance to the membrane, which depends on the diffusion coefficient of the particles. Since smaller particles contain higher diffusion coefficients than larger ones, smaller particles accumulate at higher distances away from the membrane than larger particles. The laminar flow transports particles to the detector. The closer particles accumulate toward the middle of the flow channel, the faster they are transported to the detector. As a result, smaller particles arrive earlier at the detector than larger ones. For the detection a single detector or a combination of detectors can be used. The following detectors are available: absorption detector (diode array), fluorescence detector, scattering light detector, and atom spectroscopic detector (e.g., inductively coupled plasma mass spectrometer, ICP-MS). The usage of detectors allowing deciphering the chemical composition of probes (HPLC, Raman, TXRF) has been reported [63].

Typically, polydisperse samples containing three or more different components can be separated. The injection volumes depend on the type of sample and the concentration of its particles; it may vary from between 20 μl to 2 ml. To set up a successful separation composition, concentration and pH of eluents need to be optimized to prevent aggregation or irreversible binding of the particles to the membrane [64]. During separation the particles are regularly diluted 100- to 1000-fold.

4 Conclusion

EVs can be considered as extracellular signal organelles which mediate intercellular communications. Accordingly, they are essentially involved in normal physiological and pathophysiological processes. In addition to its basic scientific importance, the young field of EV research offers an extremely high innovation potential for novel diagnostic and therapeutic procedures. Although the number of EV publications has tremendously increased in recent years, progress in the field of EV research is limited by the lack of standardized methods for their analyses as well as for their processing. Interdisciplinary collaborations of device developers and companies with scientists will certainly help to overcome these limitations within the next few years and surely will give new impacts on the exciting field of EV research.

References

1. Witwer KW, Buzas EI, Bemis LT, Bora A, Lasser C, Lotvall J, Nolte-'t Hoen EN, Piper MG, Sivaraman S, Skog J, Thery C, Wauben MH, Hochberg F (2013) Standardization of sample collection, isolation and analysis methods in extracellular vesicle research. J Extracell Vesicles 2:PMID: 24009894. doi:10.3402/jev.v2i0.20360
2. Ludwig AK, Giebel B (2012) Exosomes: small vesicles participating in intercellular communication. Int J Biochem Cell Biol 44(1):11–15. doi:10.1016/j.biocel.2011.10.005
3. Raposo G, Stoorvogel W (2013) Extracellular vesicles: exosomes, microvesicles, and friends. J Cell Biol 200(4):373–383. doi:10.1083/jcb.201211138
4. Harding C, Heuser J, Stahl P (1984) Endocytosis and intracellular processing of transferrin and colloidal gold-transferrin in rat reticulocytes: demonstration of a pathway for receptor shedding. Eur J Cell Biol 35(2):256–263
5. Harding C, Heuser J, Stahl P (1983) Receptor-mediated endocytosis of transferrin and recycling of the transferrin receptor in rat reticulocytes. J Cell Biol 97(2):329–339
6. Valadi H, Ekstrom K, Bossios A, Sjostrand M, Lee JJ, Lotvall JO (2007) Exosome-mediated transfer of mRNAs and microRNAs is a novel mechanism of genetic exchange between cells. Nat Cell Biol 9(6):654–659. doi:10.1038/ncb1596, ncb1596 [pii]
7. Pegtel DM, Cosmopoulos K, Thorley-Lawson DA, van Eijndhoven MA, Hopmans ES, Lindenberg JL, de Gruijl TD, Wurdinger T, Middeldorp JM (2010) Functional delivery of viral miRNAs via exosomes. Proc Natl Acad Sci U S A 107(14):6328–6333. doi:10.1073/pnas.0914843107, 0914843107 [pii]
8. El Andaloussi S, Mager I, Breakefield XO, Wood MJ (2013) Extracellular vesicles: biology and emerging therapeutic opportunities. Nat Rev Drug Discov 12(5):347–357. doi:10.1038/nrd3978, nrd3978 [pii]
9. Lener T, Gimona M, Aigner L, Borger V, Buzas E, Camussi G, Chaput N, Chatterjee D, Court FA, Del Portillo HA, O'Driscoll L, Fais S, Falcon-Perez JM, Felderhoff-Mueser U, Fraile L, Gho YS, Gorgens A, Gupta RC, Hendrix A, Hermann DM, Hill AF, Hochberg F, Horn PA, de Kleijn D, Kordelas L, Kramer BW, Kramer-Albers EM, Laner-Plamberger S, Laitinen S, Leonardi T, Lorenowicz MJ, Lim SK, Lotvall J, Maguire CA, Marcilla A, Nazarenko I, Ochiya T, Patel T, Pedersen S, Pocsfalvi G, Pluchino S, Quesenberry P, Reischl IG, Rivera FJ, Sanzenbacher R, Schallmoser K, Slaper-Cortenbach I, Strunk D, Tonn T, Vader P, van Balkom BW, Wauben M, Andaloussi SE, Thery C, Rohde E, Giebel B (2015) Applying extracellular vesicles based therapeutics in clinical trials – an ISEV position paper. J Extracell Vesicles 4:30087. doi:10.3402/jev.v4.30087
10. Sokolova V, Ludwig AK, Hornung S, Rotan O, Horn PA, Epple M, Giebel B (2011) Characterisation of exosomes derived from human cells by nanoparticle tracking analysis and scanning electron microscopy. Colloids Surf B Biointerfaces 87(1):146–150. doi:10.1016/j.colsurfb.2011.05.013, S0927-7765(11)00272-4 [pii]
11. Dragovic RA, Gardiner C, Brooks AS, Tannetta DS, Ferguson DJ, Hole P, Carr B, Redman CW, Harris AL, Dobson PJ, Harrison P, Sargent

IL (2011) Sizing and phenotyping of cellular vesicles using nanoparticle tracking analysis. Nanomedicine 7(6):780–788. doi:10.1016/j.nano.2011.04.003

12. Raposo G, Nijman HW, Stoorvogel W, Liejendekker R, Harding CV, Melief CJ, Geuze HJ (1996) B Lymphocytes secrete antigen-presenting vesicles. J Exp Med 183(3):1161–1172
13. Klar TA, Hell SW (1999) Subdiffraction resolution in far-field fluorescence microscopy. Opt Lett 24(14):954–956
14. Betzig E, Patterson GH, Sougrat R, Lindwasser OW, Olenych S, Bonifacino JS, Davidson MW, Lippincott-Schwartz J, Hess HF (2006) Imaging intracellular fluorescent proteins at nanometer resolution. Science 313(5793):1642–1645. doi:10.1126/science.1127344
15. Hess ST, Girirajan TP, Mason MD (2006) Ultra-high resolution imaging by fluorescence photoactivation localization microscopy. Biophys J 91(11):4258–4272. doi:10.1529/biophysj.106.091116
16. Rust MJ, Bates M, Zhuang X (2006) Sub-diffraction-limit imaging by stochastic optical reconstruction microscopy (STORM). Nat Methods 3(10):793–795. doi:10.1038/nmeth929
17. Grigorieff N, Harrison SC (2011) Near-atomic resolution reconstructions of icosahedral viruses from electron cryo-microscopy. Curr Opin Struct Biol 21(2):265–273. doi:10.1016/j.sbi.2011.01.008
18. Yuana Y, Koning RI, Kuil ME, Rensen PC, Koster AJ, Bertina RM, Osanto S (2013) Cryo-electron microscopy of extracellular vesicles in fresh plasma. J Extracell Vesicles 2:PMID:24455109. doi:10.3402/jev.v2i0.21494
19. Afework A, Beynon MD, Bustamante F, Cho S, Demarzo A, Ferreira R, Miller R, Silberman M, Saltz J, Sussman A, Tsang H (1998) Digital dynamic telepathology—the virtual microscope. Proceedings/AMIA annual symposium, pp 912–916
20. Grasselli JG, Bulkin BJ (1991) Analytical Raman spectroscopy. Wiley, New York, NY
21. Seinfeld JH, Pandis SN (2006) Atmospheric chemistry and physics: from air pollution to climate change. Wiley, New York, NY
22. Bohren C, Huffman DR (1998) Absorption and scattering of light by small particles. Wiley Science Paperback Series, New York, NY
23. Van De Hulst HC (1982) Light scattering by small particles. Peter Smith Publisher Incorporated, Gloucester, MA
24. Affolabi D, Torrea G, Odoun M, Senou N, Ali Ligali M, Anagonou S, Van Deun A (2010) Comparison of two LED fluorescence microscopy build-on modules for acid-fast smear microscopy. Int J Tuberc Lung Dis 14(2):160–164
25. Affatato S, Spinelli M, Squarzoni S, Traina F, Toni A (2009) Mixing and matching in ceramic-on-metal hip arthroplasty: an in-vitro hip simulator study. J Biomech 42(15):2439–2446. doi:10.1016/j.jbiomech.2009.07.031
26. Appidi JR, Grierson DS, Afolayan AJ (2008) Foliar micromorphology of Hermannia icana Cav. Pak J Biol Sci 11(16):2023–2027
27. Ashafa AO, Grierson DS, Afolayan AJ (2008) Foliar micromorphology of Felicia muricata Thunb., a South African medicinal plant. Pak J Biol Sci 11(13):1713–1717
28. Rigon A, Soda P, Zennaro D, Iannello G, Afeltra A (2007) Indirect immunofluorescence in autoimmune diseases: assessment of digital images for diagnostic purpose. Cytometry B Clin Cytom 72(6):472–477. doi:10.1002/cyto.b.20356
29. Brown RG (1987) Dynamic light scattering using monomode optical fibers. Appl Optics 26(22):4846–4851. doi:10.1364/AO.26.004846
30. Afaj AH, Sultan MA (2005) Mineralogical composition of the urinary stones from different provinces in Iraq. ScientificWorldJournal 5:24–38. doi:10.1100/tsw.2005.2
31. Filipe V, Hawe A, Jiskoot W (2010) Critical evaluation of Nanoparticle Tracking Analysis (NTA) by NanoSight for the measurement of nanoparticles and protein aggregates. Pharm Res 27(5):796–810. doi:10.1007/s11095-010-0073-2
32. Phillips JB, Smit X, De Zoysa N, Afoke A, Brown RA (2004) Peripheral nerves in the rat exhibit localized heterogeneity of tensile properties during limb movement. J Physiol 557(Pt 3):879–887. doi:10.1113/jphysiol.2004.061804
33. Afonso C, Fenselau C (2003) Use of bioactive glass slides for matrix-assisted laser desorption/ionization analysis: application to microorganisms. Anal Chem 75(3):694–697
34. Carmanchahi PD, Aldana Marcos HJ, Ferrari CC, Affanni JM (1998) A simple method for taking photographs of histological sections without using neither photographic camera nor microscope. BioCell 22(3):207–210
35. Konokhova AI, Yurkin MA, Moskalensky AE, Chernyshev AV, Tsvetovskaya GA, Chikova ED, Maltsev VP (2012) Light-scattering flow cytometry for identification and characterization of blood microparticles. J Biomed Opt 17(5):057006. doi:10.1117/1.JBO.17.5.057006
36. van der Pol E, Hoekstra AG, Sturk A, Otto C, van Leeuwen TG, Nieuwland R (2010)

Optical and non-optical methods for detection and characterization of microparticles and exosomes. J Thromb Haemost 8(12):2596–2607. doi:10.1111/j.1538-7836.2010.04074.x

37. Lacroix R, Robert S, Poncelet P, Kasthuri RS, Key NS, Dignat-George F, Workshop IS (2010) Standardization of platelet-derived microparticle enumeration by flow cytometry with calibrated beads: results of the International Society on Thrombosis and Haemostasis SSC Collaborative workshop. J Thromb Haemost 8(11):2571–2574. doi:10.1111/j.1538-7836.2010.04047.x
38. Orozco AF, Lewis DE (2010) Flow cytometric analysis of circulating microparticles in plasma. Cytometry A 77(6):502–514. doi:10.1002/cyto.a.20886
39. Freyssinet JM, Toti F (2010) Membrane microparticle determination: at least seeing what's being sized! J Thromb Haemost 8(2):311–314. doi:10.1111/j.1538-7836.2009.03679.x
40. Clayton A, Court J, Navabi H, Adams M, Mason MD, Hobot JA, Newman GR, Jasani B (2001) Analysis of antigen presenting cell derived exosomes, based on immuno-magnetic isolation and flow cytometry. J Immunol Methods 247(1-2):163–174, doi:S0022-1759(00)00321-5 [pii]
41. Thery C, Amigorena S, Raposo G, Clayton A (2006) Isolation and characterization of exosomes from cell culture supernatants and biological fluids. Curr Protoc Cell Biol Chapter 3:Unit 22. doi:10.1002/0471143030.cb0322s30
42. Caby MP, Lankar D, Vincendeau-Scherrer C, Raposo G, Bonnerot C (2005) Exosomal-like vesicles are present in human blood plasma. Int Immunol 17(7):879–887. doi:10.1093/intimm/dxh267, dxh267 [pii]
43. Rabesandratana H, Toutant JP, Reggio H, Vidal M (1998) Decay-accelerating factor (CD55) and membrane inhibitor of reactive lysis (CD59) are released within exosomes during In vitro maturation of reticulocytes. Blood 91(7):2573–2580
44. Steen HB (2004) Flow cytometer for measurement of the light scattering of viral and other submicroscopic particles. Cytometry A 57(2):94–99. doi:10.1002/cyto.a.10115
45. Hercher M, Mueller W, Shapiro HM (1979) Detection and discrimination of individual viruses by flow cytometry. J Histochem Cytochem 27(1):350–352
46. Nolte-'t Hoen EN, van der Vlist EJ, Aalberts M, Mertens HC, Bosch BJ, Bartelink W, Mastrobattista E, van Gaal EV, Stoorvogel W, Arkesteijn GJ, Wauben MH (2012) Quantitative and qualitative flow cytometric analysis of nanosized cell-derived membrane vesicles. Nanomedicine 8(5):712–720. doi:10.1016/j.nano.2011.09.006
47. van der Vlist EJ, Nolte-'t Hoen EN, Stoorvogel W, Arkesteijn GJ, Wauben MH (2012) Fluorescent labeling of nano-sized vesicles released by cells and subsequent quantitative and qualitative analysis by high-resolution flow cytometry. Nat Protoc 7(7):1311–1326. doi:10.1038/nprot.2012.065
48. van der Pol E, van Gemert MJ, Sturk A, Nieuwland R, van Leeuwen TG (2012) Single vs. swarm detection of microparticles and exosomes by flow cytometry. J Thromb Haemost 10(5):919–930. doi:10.1111/j.1538-7836.2012.04683.x
49. Tatischeff I, Larquet E, Falcon-Perez JM, Turpin PY, Kruglik SG (2012) Fast characterisation of cell-derived extracellular vesicles by nanoparticles tracking analysis, cryo-electron microscopy, and Raman tweezers microspectroscopy. J Extracell Vesicles 1:PMCID:PMC3760651. doi:10.3402/jev.v1i0.19179
50. Deckert-Gaudig T, Deckert V (2011) Nanoscale structural analysis using tip-enhanced Raman spectroscopy. Curr Opin Chem Biol 15(5):719–724. doi:10.1016/j.cbpa.2011.06.020
51. Binnig G, Quate CF, Gerber C (1986) Atomic force microscope. Phys Rev Lett 56(9):930–933
52. Yuana Y, Oosterkamp TH, Bahatyrova S, Ashcroft B, Garcia Rodriguez P, Bertina RM, Osanto S (2010) Atomic force microscopy: a novel approach to the detection of nanosized blood microparticles. J Thromb Haemost 8(2):315–323. doi:10.1111/j.1538-7836.2009.03654.x
53. Ashcroft BA, de Sonneville J, Yuana Y, Osanto S, Bertina R, Kuil ME, Oosterkamp TH (2012) Determination of the size distribution of blood microparticles directly in plasma using atomic force microscopy and microfluidics. Biomed Microdevices 14(4):641–649. doi:10.1007/s10544-012-9642-y
54. Sharma S, Rasool HI, Palanisamy V, Mathisen C, Schmidt M, Wong DT, Gimzewski JK (2010) Structural-mechanical characterization of nanoparticle exosomes in human saliva, using correlative AFM, FESEM, and force spectroscopy. ACS Nano 4(4):1921–1926. doi:10.1021/nn901824n
55. Palanisamy V, Sharma S, Deshpande A, Zhou H, Gimzewski J, Wong DT (2010) Nanostructural and transcriptomic analyses of human saliva derived exosomes. PLoS One 5(1), e8577. doi:10.1371/journal.pone.0008577
56. Syvitski JPM (ed) (1991) Principles, methods and application of particle size analysis. Cambridge University Press, Cambridge
57. Vogel R, Willmott G, Kozak D, Roberts GS, Anderson W, Groenewegen L, Glossop B, Barnett A, Turner A, Trau M (2011) Quantitative sizing of nano/microparticles with a tunable elastomeric pore sensor. Anal Chem 83(9):3499–3506. doi:10.1021/ac200195n

58. Afanassiev V, Hanemann V, Wolfl S (2000) Preparation of DNA and protein micro arrays on glass slides coated with an agarose film. Nucleic Acids Res 28(12), E66
59. Ito T, Sun L, Henriquez RR, Crooks RM (2004) A carbon nanotube-based coulter nanoparticle counter. Acc Chem Res 37(12):937–945. doi:10.1021/ar040108+
60. Citkowicz A, Petry H, Harkins RN, Ast O, Cashion L, Goldmann C, Bringmann P, Plummer K, Larsen BR (2008) Characterization of virus-like particle assembly for DNA delivery using asymmetrical flow field-flow fractionation and light scattering. Anal Biochem 376(2):163–172. doi:10.1016/j.ab.2008.02.011
61. Kang D, Oh S, Ahn SM, Lee BH, Moon MH (2008) Proteomic analysis of exosomes from human neural stem cells by flow field-flow fractionation and nanoflow liquid chromatography-tandem mass spectrometry. J Proteome Res 7(8):3475–3480. doi:10.1021/pr800225z
62. Zak KP, Filatova RS, Afanasyeva VV (1983) Cytochemical and ultracytochemical studies of peroxidase activity in rabbit blood granulocytes under hydrocortisone effect. Folia Haematol 110(4):490–502
63. Afifi F, Peignoux M, Auclair C (1980) Modifications of bile secretion and liver microsomal enzymes by aldosterone and spironolactone. Hepatogastroenterology 27(1):9–16
64. Baalousha M, Stolpe B, Lead JR (2011) Flow field-flow fractionation for the analysis and characterization of natural colloids and manufactured nanoparticles in environmental systems: a critical review. J Chromatogr A 1218(27):4078–4103. doi:10.1016/j.chroma.2011.04.063

Chapter 2

Tunable Resistive Pulse Sensing for the Characterization of Extracellular Vesicles

Sybren L.N. Maas, Marike L.D. Broekman, and Jeroen de Vrij

Abstract

Accurate characterization of extracellular vesicles (EVs), including exosomes and microvesicles, is essential to obtain further knowledge on the biological relevance of EVs. Tunable resistive pulse sensing (tRPS) has shown promise as a method for single particle-based quantification and size profiling of EVs. Here, we describe the technical background of tRPS and its applications for EV characterization. Besides the standard protocol, we describe an alternative protocol, in which samples are spiked with polystyrene beads of known size and concentration. This alternative protocol can be used to overcome some of the challenges of direct EV characterization in biological fluids.

Key words Extracellular vesicles, Exosomes, Microvesicles, Characterization, Quantification, Size distribution, qNano, Resistive pulse sensing

1 Introduction

Due to their small size (50–1000 nm), accurate characterization of extracellular vesicles (EVs) is technically challenging. Over time, different techniques have been developed to overcome these challenges. Most of these techniques are based on bulk analysis of EVs. For instance by total protein quantification, western blotting, bead-based flow cytometry [1] or modified protein microarrays [2]. However, alternative techniques, that allow for single particle analysis of EVs, have become recently available [3–8]. One of those techniques, provided by the qNano platform (Izon Science Ltd), is tunable resistive pulse sensing (tRPS) (Fig. 1).

In tRPS, a non-conductive membrane ("nanopore") separates two fluid cells [9] (Fig. 2). This nanopore is punctured to create a single conical shaped opening (Fig. 2, top-left). Once a voltage is applied, a current of charged ions through the nanopore is established. This baseline current is distorted, as observed by the appearance of peaks or "pulses," as particles move through the nanopore (Fig. 2, bottom). Once a particle enters the sensing zone of the

Andrew F. Hill (ed.), *Exosomes and Microvesicles: Methods and Protocols*, Methods in Molecular Biology, vol. 1545, DOI 10.1007/978-1-4939-6728-5_2, © Springer Science+Business Media LLC 2017

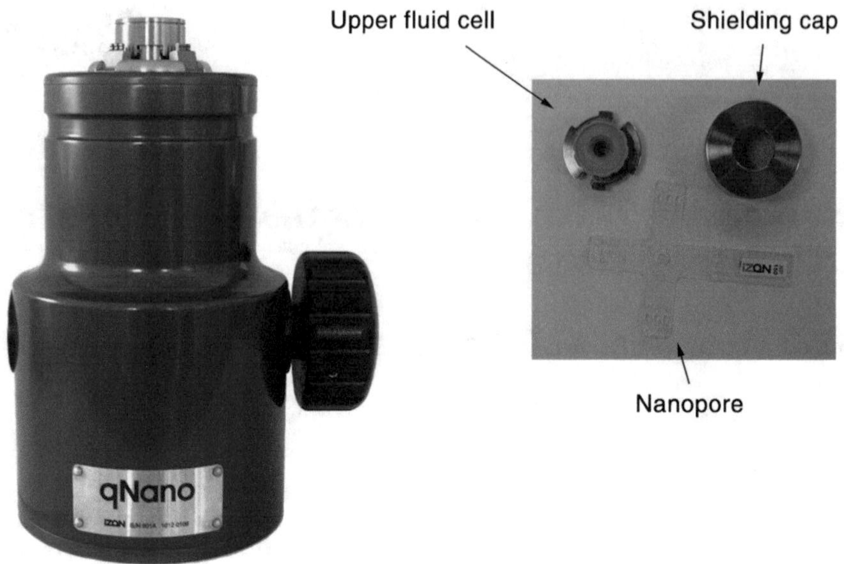

Fig. 1 Photographs of the qNano instrument and instrument parts

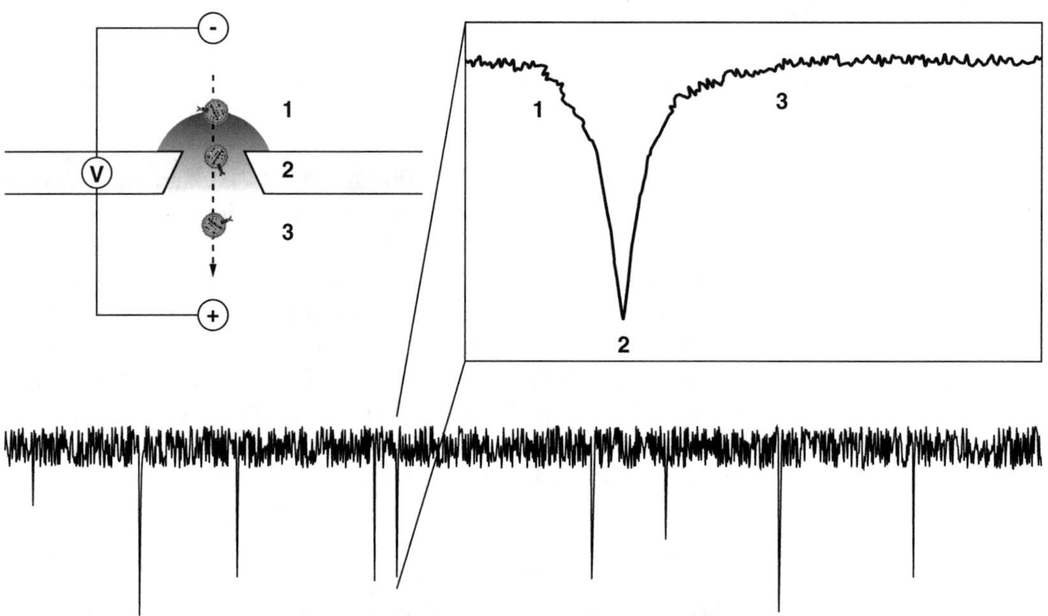

Fig. 2 The working mechanism of tunable resistive pulse sensing (tRPS). A membrane ("nanopore") with a nanosized, stretchable pore is separating two fluid compartments (top-left). After applying a voltage across the nanopore, a baseline current is established (bottom) which is disrupted by the movement of particles through the nanopore. As a particle moves towards the opening (timing 1), it starts to reduce the flow of ions through the nanopore (top-right) which will be maximum as the particle enters the nanopore opening (timing 2). This disruption reduces as the particle moves across and exits the nanopore (timing 3)

nanopore [10] (Fig. 2, timing 1), the flow of charged ions, and thus the baseline current, will be altered (Fig. 2, top-right). As the particle enters the conical opening, the relative blockade of the baseline current will be maximum (Fig. 2, timing 2). This blockade will gradually decrease to baseline levels as the particle moves further through the nanopore (Fig. 2, timing 3). To characterize particles in a sample, a calibration sample of (polystyrene) beads of known volume and concentration is measured first. The magnitude of pulses and the particle rate induced by this reference sample can subsequently be used to calculate the size profile and concentration of the particles in the measurement sample [11, 12].

The movement of particles through the nanopore is based on several independent forces, being electrokinetic (electrophoretic and electro-osmotic) and fluidic forces [10]. The variable pressure module (VPM) can be used to apply additional external force and should be used (≥0.8 kPa) to minimize interfering electrokinetic forces when analyzing particles using the smaller (NP100-NP200) nanopores [13].

Characterization of EVs using tRPS is technically challenging. Due to the heterogeneous nature of EVs a large size range of particles is usually present in a sample. Larger-sized EVs may clog the nanopore, thereby obstructing the measurement. Secondly, the sample with calibration beads should consist of the same buffer components as the EV sample. This may be technically unfeasible, as the buffer components are regularly unknown when measuring EVs, especially when measuring EVs directly in a biological sample. This problem can be overcome by using a "spiking" approach, in which the calibration beads are added to the measurement sample [3].

Here, we describe two different approaches for the characterization of EVs using tRPS. First, the standard protocol is described, which often suffices for the characterization of purified EVs. Secondly, we describe the alternative spiking approach, which could be of benefit when characterizing EVs in biological samples.

2 Materials

2.1 qNano Specific Equipment/Materials

1. qNano instrument (Izon Science Ltd, Christchurch, New Zealand).
2. Variable Pressure Module (Izon Science Ltd, Christchurch, New Zealand).
3. Polystyrene calibration particles (Izon Science Ltd, Christchurch, New Zealand) (*see* **Note 1**).
4. Nanopores (Izon Science Ltd, Christchurch, New Zealand) (*see* **Note 2**).

2.2 General Laboratory Equipment/Materials

1. Filter-tip pipette tips (*see* **Note 3**).
2. Sonication bath (*see* **Note 4**).
3. Lint-free tissues (*see* **Note 5**).
4. Phosphate buffered saline (PBS).
5. Digital calipers (supplied with the qNano instrument).

2.3 Software for Data Recording and Analysis

1. Izon Control Suite (Izon Science Ltd, Christchurch, New Zealand).
2. Spreadsheet software (*see* **Note 6**).

3 Methods

3.1 Standard Protocol

The standard protocol of tRPS-based EV quantification involves separate measurement of a (polystyrene bead-containing) calibration sample and the EV-containing sample.

1. Connect the qNano instrument to a computer running the Izon Control Suite Software. Make sure no sources of electrical interference are located close to the instrument (*see* **Note 7**).
2. Wet the lower fluid cell by introducing 75 μl PBS and immediately removing it again (*see* **Note 8**).
3. Place the nanopore of choice (*see* **Note 2**). To calibrate the stretch, use the digital calipers to measure the distance between two opposing arms of the qNano.
4. Stretch the nanopore to 47 mm and reapply 75 μl to the lower fluid cell. Prevent the formation of air bubbles in the lower fluid cell. If air bubbles are formed, remove and reapply the PBS.
5. Place the upper fluid cell and the shielding cap (which creates a "Faraday cage") on the nanopore. Add 40 μl PBS into the upper fluid cell and apply a voltage. Make sure a stable baseline current is established (*see* **Note 9**).
6. Dilute the calibration particles in PBS to the target concentration of the used nanopore (*see* **Note 10**).
7. Remove the PBS from the upper fluid cell and apply 40 μl of the calibration particles into the upper fluid cell. Make sure a stable baseline current is established (*see* **Note 9**). Reduce the applied stretch slowly towards 43 mm and observe the blockades caused by the calibration particles. Stop reducing the stretch when the mode blockade caused is at least 0.1 nA, but preferable >0.3 nA (*see* **Notes 11** and **12**).

8. Apply ≥0.8 kPa pressure using the VPM and click "record" (*see* **Note 13**). Make sure that a particle rate (*see* **Note 14**) of >100 min⁻ and a mode blockade height of >0.1 nA is recorded (*see* **Note 12**).
9. If the baseline current suddenly drops or keeps drifting during recording, pause the recording and try to reestablish a stable current (*see* **Note 9**).
10. Record >500 particles, for at least 30 s (*see* **Note 14**). Fill out the details of the calibration sample in the pop-up form.
11. Optionally, multi-pressure measurement can be performed (*see* **Notes 13** and **15**). Hereto, add at least 0.2 kPa and record a second measurement (more steps could increase accuracy).
12. Remove the calibration sample and wash the upper fluid cell by resuspending 100 μl PBS in the upper fluid cell 3–4 times. Remove residual PBS by usage of the lint-free tissue (*see* **Note 16**).
13. Introduce the EV sample and make sure the baseline current is within 3% of the baseline for the calibration sample (*see* **Note 17**).
14. Record the sample at the same VPM pressure(s) as applied for the calibration sample.
15. Click the "Analyse data" tab and right-click on "Unprocessed files" and select "Process files".
16. Click on the checkbox in the "calibrated" column next to one of the sample files. This will initialize the calibration pop-up menu. Select the "multi-pressure measurement" tab if applicable and select the sample files and calibration file(s).
17. Once calibrated, an EV sample file will display a size distribution in nm instead of nA (Fig. 3, right). Click on "Preview" to generate a .pdf file containing statistics such as the concentration (measured and raw if a diluted sample was used).

3.2 Spiking the Sample with Polystyrene Beads of Known Size and Concentration

The standard protocol for tRPS-based EV quantification relies on usage of appropriately formulated calibration samples (i.e., with the diluents resembling the fluid of the EV sample). This may be unfeasible for biological fluids, since their exact composition may be unknown rendering their simulation impossible. Secondly, the volume of the biological sample (e.g., only 100 μl of plasma) may be insufficient for preparation of calibration fluid (which usually can be done by removal of small particulate matter by ultracentrifugation or filtering). In such cases, an alternative is provided by performing a spiking protocol, in which calibration beads are introduced in the EV sample [3]. This methodology can also be used when samples are measured over a prolonged period of time and stable nanopore conditions cannot be guaranteed due to nanopore clogging.

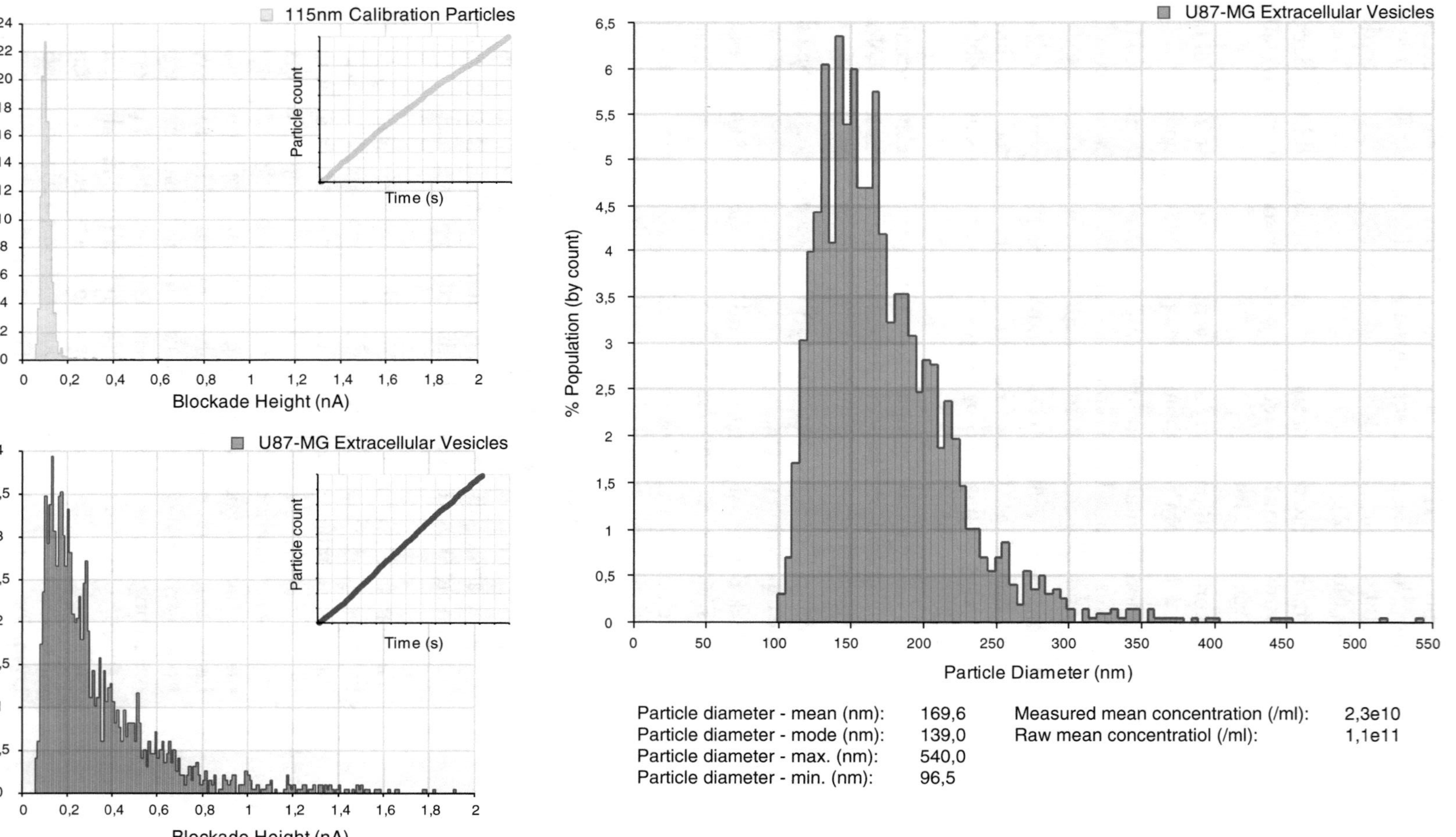

Fig. 3 Characterization of extracellular vesicles (EVs) by calibrating the EVs to polystyrene beads of known size and concentration. A sample of 115 nm polystyrene beads is measured, resulting in a size distribution and particle rate plot (top-left). Secondly, the EVs, purified by sequential ultracentrifugation, are measured resulting in a second size distribution and particle rate (bottom-left). Once the EV sample is calibrated to the reference beads, the recorded blockades in nA, can be calculated to absolute sizes in nm (right-side). The particle rates are used to calculate the concentration of the EVs

1. Setup the qNano instrument as outlined in Subheading 3.1 **steps 1–5**.
2. Check the approximate particle rate of the EV samples.
3. Dilute the EV sample in PBS (*see* **Note 18**).
4. Determine the dilution of polystyrene beads that is needed to obtain a count rate that resembles the count rate of the EV samples (*see* **Note 19**), and check for the ability to distinguish EVs and polystyrene beads (*see* **Note 20**).
5. Prepare the samples by diluting polystyrene beads into the samples (*see* **Note 21**). Also prepare a "beads-only" sample (*see* **Note 22**).
6. Record the beads-only and sample measurements, preferable in triplicate (*see* **Note 23**).
7. Process all files as outlined in Subheading 3.1 **step 15**.
8. Display the size distribution graphs (uncalibrated) of the beads-only samples and sample files (Fig. 4, left). Determine at which nA value a cutoff can be set to distinguish the two populations (Fig. 4) (*see* **Note 24**).
9. Obtain the total particle count (in sample details window) for each sample and put this into a spreadsheet software program (Table 1).
10. Click the "filter options" button to obtain the filter settings. Enter the cutoff obtained in **step 8** and filter the samples. Make sure to select the "apply to all samples in group" checkbox to filter all samples directly.

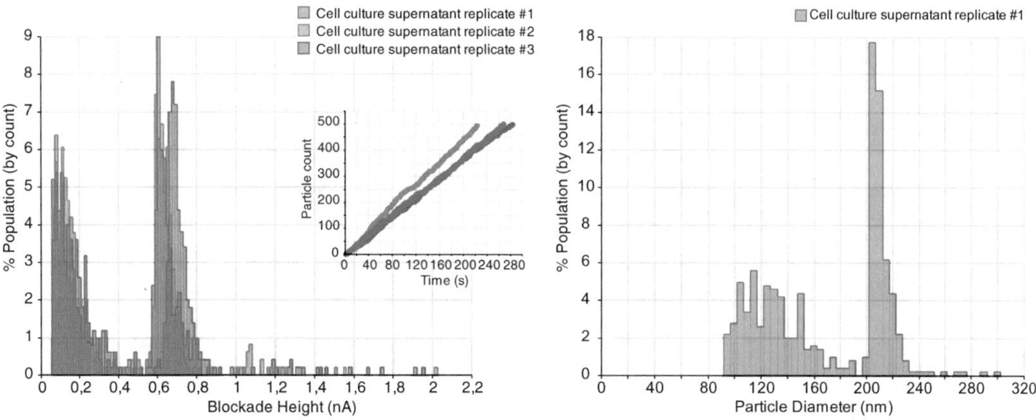

Fig. 4 Quantification and size estimation of EVs by spiking the sample with polystyrene beads of known size and concentration. Three replicates of glioblastoma cell culture supernatant spiked with 203 nm polystyrene beads are measured (left). All particles smaller than 0.48 nA were determined EVs. The EV-to-beads ratio is used to calculate the concentration of EVs. The spiked polystyrene beads can be used to obtain an accurate size distribution without the need of an external calibration sample (right)

Table 1
Example calculation of EV concentration using the alternative spiking method

Sample	Beads-only #1	Beads-only #2	Replicate #1	Replicate #2	Replicate #3
Average current	63	60	62	61	61
Rate	57	74	112	133	105
Cutoff used	0.48	0.48	0.48	0.48	0.48
Total particles	297	299	502	495	500
Extracellular vesicles (EVs)	27	46	254	240	246
Beads + multimers	270	253	248	255	254
EVs/beads	0.100	0.182	1.024	0.941	0.969
Sample—background			0.88	0.80	0.83
EVs (10^8/ml) in sample			8.83	8.00	8.28
Dilution factor of EVs			2.5	2.5	2.5
EVs (10^8/ml) raw			22.08	20.01	20.69

11. Obtain the total particle counts for each sample after the filter step. Fill out these numbers into the spreadsheet software (*see* Table 1 for an example calculation).
12. Subtract the EV counts from the total counts to obtain the amount of calibration particles. Subsequently, divide the number of EVs by the number of polystyrene particles to obtain the EV-to-bead ratio.
13. To account for background particles, subtract the average ratio obtained for the beads-only samples from each EV-to-bead ratio.
14. Multiply the EV-to-bead ratio to the concentration of polystyrene beads in the sample. Secondly, multiply this value by the dilution factor of the EVs (*see* **Note 25**) to obtain the raw concentration of EVs.
15. Optionally: introduce a correction when overlap of EVs and polystyrene beads is observed (*see* **Note 26**).

3.3 Obtaining an EV Size Distribution from a Spiked Sample

The above-described spiking procedure can also be utilized to obtain a proper size distribution profile of EVs in case the preparation of appropriate calibration samples is impossible.

1. Prepare, measure, and process the EV samples as outlined in Subheading 3.2 **steps 1–8**.
2. Once processed open the sample of interest twice in the Izon Control Suite.

3. For one of the files, filter the sample to display particles larger than the determined cutoff only. Set this sample as "calibration" and enter the mode size of the calibration particles.
4. Couple the sample file and the newly create calibration file as outlined in Subheading 3.1 **step 16**.
5. Once successfully coupled, the unknown sample can now be displayed as a size distribution in nm based on the spiked calibration particles (Fig. 4, right). This graph will display two populations, one for the EVs and one for the reference particles.

4 Notes

1. For EV characterization different polystyrene beads are used: CPC100, CPC200, and CPC400, with mode diameters of 115, 203, and 335 nm, respectively (these numbers may vary based on the batch used).
2. Different sizes of nanopores are used: the NP100 nanopore (optimal size range 70–200 nm), NP150 (80–300 nm), and NP200 (100–400 nm). Due to heterogeneity of EV samples, the NP150 and NP200 are most often used for characterization of EVs.
3. To minimize background particle detection, we use filter-tip pipette tips.
4. To homogenize the calibration particles a basic tabletop sonicator can be used.
5. To completely remove any residual liquids between measurements, lint-free tissue can be used. To minimize contamination of background particles, lint-free tissue is preferred over regular tissues.
6. For almost all data analyses the Izon Control Suite can be used. However, all data-points can be exported for analysis in other software packages. For EV quantification using the spiking method, a spreadsheet software program is required.
7. Electronic devices used in close proximity of the instrument can significantly interfere with the detection signal. This interference is observed as identical, quickly repeating short pulses. We have most often observed this interference caused by mobile phones.
8. This is done to decrease the risk of air-bubble formation in the lower fluid cell. Air bubbles can be a major source of instable baseline current.
9. The baseline current depends on the applied buffer, stretch and voltage. The current should be stable and the root mean square

(RMS) noise should be <10 pA. If these conditions are not met, air bubbles or (partial) nanopore blocking may be causative. To solve this, resuspend the sample in the upper fluid cell and check if the baseline becomes stable. If not, remove both the sample and the PBS in the lower fluid cell. If (after reapplication of PBS) no stable baseline current is established, the nanopore may be (partially) blocked. Tap the shielding cap (using the supplied plunger) to vibrate the nanopore and to disrupt particles. Clogging may also be solved by induction of a brief pressure by pushing down and pulling out of the plunger. Alternatively, the shielding cap can be put in place whilst pressing on the nanopore, which will vibrate the nanopore. Also, the nanopore can be maximally stretched (i.e., 47 mm), combined with applying maximal external pressure. If still unsuccessful, remove the nanopore and rinse heavily using deionized water. Re-place the nanopore on the instrument.

10. Each nanopore has a target concentration. For the NP100 and NP150 nanopores the target concentration is 10E10 per ml and for the NP200 the target concentration is 10E9 per ml.

11. The blockade height caused by a particle moving through the nanopore is based on the stretch, the applied voltage and the buffer used. If the nanopore opening is reduced (less stretch applied) the relative blockade by the particle will increase. This also implies that smaller particles surpass the detection threshold. Larger particles, on the other hand, will block the nanopore more frequently. By increasing the voltage applied, the flow of ions will increase and so will the (relative) blockade caused by particles moving through the nanopore. However, an increased voltage can also result in increased RMS noise. The flow of ions, and thus a higher baseline current, can also be established by using a buffer with increased salt concentration. However, this may influence the EV characteristics, for instance caused by changes in osmosis.

12. For accurate detection of particles a mode blockade of at least 0.1 nA is required. However, the mode blockade set for the calibration particles will also determine the range of EVs detectable by the instrument. For instance, a mode blockade of 0.1 nA for 203 nm calibration beads indicates that the instrument will only be able to detect particles that are slightly smaller than 203 nm. Reducing the stretch or increasing the voltage (*see* **Note 11**) could be needed to decrease the lower detection limit.

13. External pressure needs to be applied to counteract the influence of electrokinetic forces. These electrokinetic forces are not negligible when using small pore sizes (NP100-NP200) [14], which is often the case upon EV quantification. Since EVs display a modest zeta potential (i.e., the potential

difference between the dispersion medium and the stationary layer of fluid attached to the particle) [15, 16] the influence of the electrokinetic forces is low and can be completely abolished by applying >0.8 kPa external pressure [13].

14. The particle rate recorded (particles per minute) will depend on the concentration of the particles, the applied pressure and applied stretch (the rate will decrease by decreasing the stretch). Since at least 500 particles should be recorded, a particle rate of >100 per minute is advised but not required. In our experience particle rates >2000 per minute will be less reliable.
15. Multi-pressure measurement is advisable when measuring EVs with increased surface charge (e.g., as a result of coupling highly charged ligands to the surface). In such cases, difference in surface charge between EVs and polystyrene calibration beads will result in inaccurate concentration estimations as one of the particle sets is more likely to move through the nanopore than the other [14]. Measurement of the calibration bead sample and EV sample at multiple pressures provides additional data that is used to accurately calculate the concentration of EVs.
16. Residual PBS in the upper fluid cell can dilute the measurement sample. To prevent this, remove the upper fluid cell and gently wipe lint-free tissue in the bottom-opening of the cell.
17. To accurately compare a calibration sample with an EV sample, the baseline current should not differ more than 3%. If unable to reach a comparable baseline current, apply the strategy outlined in Note 9. Alternatively, dilution of the sample in PBS could make the EV sample more comparable to the calibration sample.
18. Dilution in PBS may facilitate EV counting by the qNano instrument. However, to guarantee appropriate counting of EVs, try to keep the particle rate above 70–100 particles per minute (*see* **Note 14**).
19. Although not strictly necessary, an EV-to-bead ratio of approximately 1 will make the measurements most reliable. If EVs or beads outnumber their counterparts the calculation of concentrations will be more prone to variation.
20. To distinguish EVs from polystyrene beads, both populations should be identifiable based on blockade sizes. For EV quantifications in biological samples we tend to use a NP200 nanopore in combination with CPC400 (mode 335 nm) polystyrene beads or an NP150 nanopore in combination with 203 nm beads. To maximize the population of EVs detected, try to obtain settings where the polystyrene beads induce blockades of at least 0.5 nA. By increasing the blockade height caused by the polystyrene particles, the detection limit for the EVs will decrease (*see* **Note 12**).

21. Example sample preparation:
    (a) 40 μl cell culture supernatant (after 5 min 300×*g* centrifugation to remove cells).
    (b) 40 μl PBS.
    (c) 20 μl 1:200 diluted 203 nm polystyrene beads (stock 1e12 ml^{-1}).
22. A beads-only sample is used to quantify background particles and to identify the population of polystyrene beads. For this sample "EV free cell culture medium" should be used that has received the same treatments as the samples of interest, but lacks EVs.
23. To spread variation in nanopore conditions, each set of samples should be measured once before recording duplicates and triplicates. Prepare fresh samples (i.e., addition of PBS and beads) directly before each measurement.
24. Setting the cutoff remains arbitrary. Make sure each sample has the same bin-size setting (ViewSettings panel, accessible by clicking the popup button in the View panel). We choose to set the cutoff at 0.48 nA (Fig. 4, left). All particles smaller than the cutoff are determined EVs.
25. Since the EVs are diluted (upon mixing with calibration beads and addition of PBS) the obtained concentration should be corrected for this. For the example setup outlined in note 21, the EVs are diluted 2.5 times.
26. A correction can be introduced when the detection of EVs and polystyrene beads overlaps. Measure the EV sample without polystyrene beads and determine the "Bead-to-EV ratio" based on the cutoff determined in Subheading 3.2 **step 8** (here the term "bead" refers to the fraction of EVs that are detected within the spiked-bead-detection range). Usually this ratio is insignificant, but if not add this Bead-to-EV ratio to the EV-to-bead ratio as determined in Subheading 3.2 **step 13**. This new ratio should be used for the remaining steps in the protocol.

Acknowledgment

We thank J. Berenguer (VUmc, Amsterdam, The Netherlands) for providing us with the glioblastoma cell culture supernatant.

This work has been financially supported, in part, by the Dutch Hersenstichting (foundation concerned with diseases of the brain), the Schumacher Kramer Stichting (Foundation), and the T&P Bohnenn foundation.

References

1. Thery C, Amigorena S, Raposo G, Clayton A (2006) Isolation and characterization of exosomes from cell culture supernatants and biological fluids. Curr Protoc Cell Biol 3:22. doi:10.1002/0471143030.cb0322s30
2. Jorgensen M, Baek R, Pedersen S, Sondergaard EK, Kristensen SR, Varming K (2013) Extracellular Vesicle (EV) Array: microarray capturing of exosomes and other extracellular vesicles for multiplexed phenotyping. J Extracell Vesicles 2:PMID: 24009888. doi:10.3402/jev.v2i0.20920
3. de Vrij J, Maas SL, van Nispen M, Sena-Esteves M, Limpens RW, Koster AJ, Leenstra S, Lamfers ML, Broekman ML (2013) Quantification of nanosized extracellular membrane vesicles with scanning ion occlusion sensing. Nanomedicine 8(9):1443–1458. doi:10.2217/nnm.12.173
4. Gardiner C, Ferreira YJ, Dragovic RA, Redman CW, Sargent IL (2013) Extracellular vesicle sizing and enumeration by nanoparticle tracking analysis. J Extracell Vesicles 2:PMCID: PMC3760643. doi:10.3402/jev.v2i0.19671
5. Nolte-'t Hoen EN, van der Vlist EJ, Aalberts M, Mertens HC, Bosch BJ, Bartelink W, Mastrobattista E, van Gaal EV, Stoorvogel W, Arkesteijn GJ, Wauben MH (2012) Quantitative and qualitative flow cytometric analysis of nanosized cell-derived membrane vesicles. Nanomedicine 8(5):712–720. doi:10.1016/j.nano.2011.09.006
6. Dragovic RA, Gardiner C, Brooks AS, Tannetta DS, Ferguson DJ, Hole P, Carr B, Redman CW, Harris AL, Dobson PJ, Harrison P, Sargent IL (2011) Sizing and phenotyping of cellular vesicles using nanoparticle tracking analysis. Nanomedicine 7(6):780–788. doi:10.1016/j.nano.2011.04.003
7. van der Vlist EJ, Nolte-'t Hoen EN, Stoorvogel W, Arkesteijn GJ, Wauben MH (2012) Fluorescent labeling of nano-sized vesicles released by cells and subsequent quantitative and qualitative analysis by high-resolution flow cytometry. Nat Protoc 7(7):1311–1326. doi:10.1038/nprot.2012.065
8. Maas SLN, De Vrij J, Broekman MLD (2014) Quantification and size-profiling of extracellular vesicles using tunable resistive pulse sensing. J Vis Exp 92:e51623
9. Roberts GS, Kozak D, Anderson W, Broom MF, Vogel R, Trau M (2010) Tunable nano/micropores for particle detection and discrimination: scanning ion occlusion spectroscopy. Small 6(23):2653–2658. doi:10.1002/smll.201001129
10. Kozak D, Anderson W, Vogel R, Chen S, Antaw F, Trau M (2012) Simultaneous size and zeta-potential measurements of individual nanoparticles in dispersion using size-tunable pore sensors. ACS Nano 6(8):6990–6997. doi:10.1021/nn3020322
11. Vogel R, Willmott G, Kozak D, Roberts GS, Anderson W, Groenewegen L, Glossop B, Barnett A, Turner A, Trau M (2011) Quantitative sizing of nano/microparticles with a tunable elastomeric pore sensor. Anal Chem 83(9):3499–3506. doi:10.1021/ac200195n
12. Willmott GR, Vogel R, Yu SS, Groenewegen LG, Roberts GS, Kozak D, Anderson W, Trau M (2010) Use of tunable nanopore blockade rates to investigate colloidal dispersions. J Phys Condens Matter 22(45):454116. doi:10.1088/0953-8984/22/45/454116
13. Yang L, Broom MF, Tucker IG (2012) Characterization of a nanoparticulate drug delivery system using scanning ion occlusion sensing. Pharm Res 29(9):2578–2586. doi:10.1007/s11095-012-0788-3
14. Kozak D, Anderson W, Trau M (2012) Tuning particle velocity and measurement sensitivity by changing pore sensor dimensions. Chem Lett 41(10):1134–1136. doi:10.1246/cl.2012.1134
15. Marimpietri D, Petretto A, Raffaghello L, Pezzolo A, Gagliani C, Tacchetti C, Mauri P, Melioli G, Pistoia V (2013) Proteome profiling of neuroblastoma-derived exosomes reveal the expression of proteins potentially involved in tumor progression. PLoS One 8(9):e75054. doi:10.1371/journal.pone.0075054
16. Sokolova V, Ludwig AK, Hornung S, Rotan O, Horn PA, Epple M, Giebel B (2011) Characterisation of exosomes derived from human cells by nanoparticle tracking analysis and scanning electron microscopy. Colloids Surf B Biointerfaces 87(1):146–150. doi:10.1016/j.colsurfb.2011.05.013

Chapter 3

Immuno-characterization of Exosomes Using Nanoparticle Tracking Analysis

Kym McNicholas and Michael Z. Michael

Abstract

Due to their size, small extracellular vesicles such as exosomes have been difficult to identify and to quantify. As the roles that exosomes play in intercellular signalling become clearer, so does their potential utility as both diagnostic biomarkers for disease and as therapeutic vectors. Accurate assessment of exosomes, both their number and their cargo, is important for continued advancement in the field of vesicle research. To that end, several technologies, including nanoparticle tracking analysis, have been developed to define the physical characteristics of vesicle preparations and determine their concentration. This chapter describes a method for identifying the size and concentration of a subpopulation of vesicles in biological samples, using nanoparticle tracking analysis. Characterization of distinct exosomes is enabled by specific marker antibodies, coupled to fluorescent quantum dots.

Key words Quantum dots, Qdots®, NanoSight, CD63, Exosomes, Microvesicles

1 Introduction

Exosomes are extracellular vesicles, released from a wide variety of cell types, and are being increasingly recognized as mediators of intercellular signaling and potential biomarkers of human disease [1]. They comprise a specific subset of cell-derived vesicles, formed through the endosomal pathway [2]. The precise definition of an exosome is still open to some debate, even regarding their size. Early exosomal research was based on electron micrographs showing cup-shaped vesicles up to 100 nm in diameter, but more recent work has widened this range from 50-180 nm [3–5]. Regardless, it is their small size that poses unique challenges for their study. Nanoparticle tracking analysis (NTA) has been recently devised as a means of both quantifying and determining the size of particles between 30 nm and 1 μm in diameter. This analysis is reliant on a NanoSight (NanoSight Limited, Malvern Instruments, Amesbury,

Andrew F. Hill (ed.), *Exosomes and Microvesicles: Methods and Protocols,* Methods in Molecular Biology, vol. 1545,
DOI 10.1007/978-1-4939-6728-5_3, © Springer Science+Business Media LLC 2017

UK) instrument, which is an upright microscope fitted with a high-sensitivity video camera and a laser light source above which the sample is retained as a particulate suspension. The NanoSight Tracking Analysis software monitors the movement of particles, such as exosomes as small as 40 nm, in real time, and has been successfully used to characterize vesicle populations in a wide variety of human samples including plasma, placenta [5], serum [3], urine [6] as well as conditioned media from cell lines including human lymphoblastoid T-cells [7] and breast cancer cells [8]. In addition to sizing vesicles, NTA also addresses the issue of characterizing vesicles, by virtue of cellular markers, using antibodies conjugated to fluorescent moieties, such as quantum dots [9].

Several proteins that are located on the outer surface of exosomes have been identified to distinguish exosomes from other microvesicles. These include members of the tetraspanin protein family, CD9, CD81, and especially CD63, which are known to be highly enriched and detectable in exosome preparations [3, 10, 11]. Antibody-mediated fluorophore labeling of markers such as CD63, EpCAM [3] and NDOG2 [5] has been combined with NTA to successfully determine the size and concentration of exosomes in a sample.

The NanoSight instrument tracks vesicles by passing a laser beam through the liquid sample and detecting light scattered from nanoparticles in suspension. The light scatter is captured in a short video, which enables the Brownian motion of particles in this sample to be determined on a frame-by-frame basis. The distance moved by a particle is used in the calculation of the particle diffusion (*Dt*), a number which is then used to estimate the particle's hydrodynamic diameter using the Stokes–Einstein equation [5]. The NanoSight LM10 can be fitted with diodes for different laser wavelengths (405, 488, 532, and 635 nm). A matched long-pass filter allows the NTA software to visualize only the fluorescently labeled particles excited by the laser. In normal scattered light the total population of particles in a sample can be tracked. With the long-pass filter in place, a subset of particles labeled with the fluorescent quantum dots can be identified and measured.

This method describes, in four major procedures, the process in which a NanoSight LM10 can be used to detect and measure the size and concentration of exosomes labeled with an antibody–quantum dot conjugate in a sample. The first procedure describes the isolation of microvesicles from cell medium and biological samples by ultracentrifugation. The second step involves the conjugation of an anti-human CD63 antibody to a quantum dot using a commercially available kit. This conjugate is then incubated with a vesicle preparation and finally viewed under the NanoSight unit in both normal light scatter and fluorescent modes.

2 Materials

1. HT29 human colon adenocarcinoma cell line.
2. Dulbecco's modified Eagle's medium (DMEM) High Glucose 1× (GIBCO, Life Technologies).
3. Exo-FBS™ (System Biosciences Inc.).
4. 10× phosphate buffered saline (PBS): 80 g NaCl, 2.0 g KCl, 14.4 g Na_2HPO_4, 2.4 g KH_2PO_4. Add water to 1 L. Dilute to 1× as needed with water. Adjust pH to 7.4. Filter (0.2 μm) and autoclave. Store at room temperature.
5. Qdot® 605 Antibody Conjugation Kit (Life Technologies). Store at 4 °C.
6. Mouse monoclonal antibody to human CD63 clone H5C6 (300 μg; BD Pharmingen/Biosciences™).
7. 75 % ethanol.
8. Minisart 0.22 μm sterile filters (Sartorius).
9. Nonstick hydrophobic microcentrifuge tubes 0.65 mL.
10. Ultracentrifuge tubes.
11. Exoquick-TC™ (System Biosciences Inc.).

3 Methods

3.1 Vesicle Isolation from Conditioned Media (Adapted from [12])

Microvesicles are commonly isolated from conditioned media and biological samples, such as urine [13], by a two-step differential centrifugation process. Initially the sample is centrifuged at low speed to sediment cells, membrane fragments and other debris. The supernatant containing exosomes is then filtered and centrifuged at high speed to sediment microvesicles. A colorectal cancer cell line HT29 was cultured at 37 °C in a 5 % CO_2 humidified incubator as an adherent monolayer in DMEM media supplemented with 10 % exosome-depleted serum Exo-FBS™ (*see* **Note 1**). Conditioned medium was collected in 15 mL tubes and processed as follows:

1. Spin conditioned media at low speed (1500 ×*g*) 4 °C for 20 min.
2. Aliquot supernatant. Spin 10,000 ×*g* at 4 °C for 45 min.
3. Aliquot supernatant into ultracentrifuge tubes. Spin 150,000 ×*g* at 4 °C for 2 h.
4. Discard supernatant. Resuspend pellet in 1.5 mL 1× PBS. Mix well.
5. Filter preparation (0.22 μm).
6. Aliquot into 1.5 mL ultracentrifuge tube. Spin 110,000 ×*g* 4 °C for 70 min.

7. Discard supernatant. Resuspend vesicle pellet in 35 μL 1× PBS. Mix well. Store at −80 °C in nonstick hydrophobic microcentrifuge tubes.

3.2 Vesicle Isolation from Biofluids

Biofluids (e.g., saliva) are received fresh, stored on ice and processed on the same day.

1. Aliquot saliva into 1.5 mL microcentrifuge tube and centrifuge 8000×*g* at RT for 5 min.
2. Transfer supernatant into 15 mL tube. Add filtered (0.22 μm) 1× PBS to final 5 mL volume. Centrifuge, 1500×*g*, 4 °C for 20 min. Samples can then be treated as per the conditioned media.

3.3 Antibody Conjugation

Quantum dots are fluorescent semiconductor nanocrystals (10–20 nm diameter) that can be conjugated to a primary antibody of choice using commercially available kits, with a range of emission wavelengths (*see* **Note 2**).

1. 300 μg of mouse monoclonal anti-human CD63 (BD Biosciences™) antibody was conjugated to quantum dots, with an emission maximum of 605 nm, according to the manufacturer's instructions (Qdot® 605 Antibody Conjugation Kit, Life Technologies) and stored at 4 °C, protected from light (*see* **Note 3**).
2. As a negative control, mouse immunoglobulin G has been labeled in the same way (3). Most commercial Qdot conjugation kits include steps for the removal of excess antibody and unbound Qdots. Unconjugated reagent can give misleading NTA results, so should be removed. Washing the sample through a filter, such as a 300 kDa centrifugal filter at low speed, has been suggested [14] [5].

3.4 Immunofluorescence Labeling with Antibody–Quantum Dot Conjugate

The optimum concentration of the antibody–Qdot conjugate is dependent on the vesicle concentration and the binding efficiency of the conjugate.

1. Efficient labeling was achieved by overnight incubation of the CD63-Qdot conjugate with a high concentration of exosomes followed by dilution of sample and removal of unbound Qdots.
2. The conjugate was added to give a final concentration of 10 nM. Incubation times should be optimized and can extend from 15 min [5], or 2 h [3] at room temperature, to overnight incubation at 4 °C.

3.5 Removal of Unbound Antibody–Quantum Dot Conjugates

Following incubation of the vesicle preparation with the antibody conjugated Qdots, unbound Qdots were removed prior to detection. After an overnight incubation, diluting the sample with 1× PBS and further precipitating bound exosomes with Exoquick-TC™

(System Biosciences) greatly reduces the background caused by unbound Qdots.

1. Add one volume Exoquick-TC™ to five volumes sample (or according to the manufacturer's instructions), mix well by pipetting and incubate overnight at 4 °C.
2. Following incubation, centrifuge at 1500 ×*g*, for 45 min at 4 °C.
3. Resuspend vesicle pellet in 1× PBS prior to further dilution for analysis on the NanoSight LM10.

3.6 Loading a Sample into the NanoSight Chamber

The following steps apply to a NanoSight LM10 fitted with a 405 nm laser (blue/violet), 430 nm long-pass filter and a high sensitivity digital camera system (OrcaFlash2.8, Hamamatsu C11440, NanoSight Ltd). Constant sample flow was not used (*see* **Note 4**).

1. A minimum volume of 0.25 mL can be loaded into the LM10 chamber. Slowly inject the sample containing microvesicles in suspension into the luer lock port of the chamber using a 1 mL syringe. Care should be taken to avoid introducing air bubbles into the chamber. Ensure all air bubbles are removed before switching on the laser.
2. Allow the sample to stand for a short time (up to 15 s). This avoids drift affecting the analysis of the video.
3. When the laser is on, the beam should appear as a thin blue line passing through the chamber, in normal light scatter mode. Ideally, all videos should be captured with the sample as close as possible to the "thumbprint" (flare spot where the laser beam emerges) without obvious interference from light scattered by the thumbprint itself. This is particularly important when the fluorescent filter is used. Also, multiple videos should be taken in the same position relative to the thumbprint. If the optical flat (prism) has a chrome finish, then the optimal viewing position is just after the vertical line, down beam from the thumbprint. Ensure the laser is switched off after video capture has finished.
4. Chamber temperature is recorded manually, using the provided thermometer, and ranges between 20 and 25 °C. This value is then used for temperature compensation by the NTA analysis software.
5. It is advisable to initially load only diluent into the chamber, to check the cleanliness of the diluent and the chamber. A sixty-second video of this diluent should show less than 200 completed tracks. If more are seen then replace diluent and clean the chamber and optical flat before reloading.

3.7 Analysis of Exosome Preparations with Qdot Conjugated Antibodies

1. The exosome preparation should be diluted so that the number of particles (per mL) falls between 10^8 and 10^9 and does not contain particles larger than 1 μm in diameter.
2. A minimum of three videos per sample should be captured and analyzed within a short period of time. Ideally a minimum of 1000 tracks should be captured within a 60-s video. Sample adherence to the chamber may reduce concentration measurements if it is left in the chamber for longer periods of time.
3. For consistency and accuracy, all settings should be kept constant between samples. For example, the minimum expected particle size determines the maximum distance that the software expects a particle to move from one frame to the next. Since NTA release 2.3, this value is calculated from the first ten seconds of the video, which is preferable to the operator having to predict a value. Recently, Gardiner et al. [14] have provided a useful summary of the NTA software settings and their values appropriate for the dispersion of the sample.
4. Similarly the camera level must be kept constant between samples in order to compare sample concentrations. The highest camera sensitivity (level 16) may be needed when the fluorescent filter is in place because the fluorescent signal is dimmer than the signal in light scatter mode (*see* **Note 5**). Often this entails starting with a higher concentration of sample for fluorescent mode and then diluting the sample to obtain good quality reads in normal light scatter mode. This dilution will depend on the binding efficiency of the Qdot conjugate.
5. By integrating multiple measurements (minimum $n = 3$) for each sample, in both light scatter mode and fluorescence mode, specific subsets of microparticles can be quantified (% bound vesicles) and characterized according to the external markers they possess.
6. Determine the profile of unbound detection reagent by analyzing a sample containing only antibody Qdot conjugate, and diluent, with the NanoSight instrument. With the fluorescent filter in place, finely focus the camera until the smallest particles are single points of light (*see* also **Notes 6–8**). Unbound Qdots typically appear as a peak in the range 20–40 nm.

4 Notes

1. To prevent bovine extracellular vesicles confounding measurements, cell culture media should be depleted of exosomes and other particles. Commercial exosome-depleted fetal bovine sera are available.

2. To enable experimental flexibility, Qdots can be conjugated to intermediate reagents, such as streptavidin, that enable detection using a range of biotinylated antibodies, but care should be taken to block samples that may contain endogenous biotin thereby avoiding a nonspecific signal.
3. Alternative labeling kits should be considered. Conjugation kits that employ SiteClick™ chemistry may provide more specific conjugation of the Qdot to the heavy chain of IgG antibodies, thus ensuring that conjugated antibodies can bind to their antigen.
4. While Qdots are quite stable, prolonged exposure to intense light can lead to photo-bleaching of many fluorophores. This problem may be overcome by delivering particle suspensions into the sample chamber, under constant flow. This is standard for later NanoSight models (e.g., NS300 and NS500), however the LM10 model will require a syringe pump assembly. The NTA software compensates for flow during analysis.
5. Labeled vesicles are very dim compared to light scatter mode. The brightness can be enhanced using the greyscale histogram in NTA software, as well as reducing the background.
6. Care should be taken when viewing preparations of cell culture media containing phenol red, such as DMEM. This medium will naturally fluoresce under the laser, potentially yielding misleading results. To check for this, load unlabeled sample and check that the sample does not fluoresce in fluorescence mode.
7. If necessary, silica or fluorescent microspheres (Polysciences Inc., Warrington, PA, USA) may be used as standards to calibrate the NanoSight instrument.
8. Clean the chamber with diluent between samples. To avoid cross-contamination, diluent and then compressed air should be passed through the fittings. The optical flat should be removed and thoroughly cleaned with 70% ethanol.

Acknowledgments

Thanks to Rebecca Dragovic and NanoSight representatives for helpful discussions. Projects related to this chapter were funded by the Flinders Medical Centre Foundation and BioInnovation SA.

References

1. Skog J, Würdinger T, van Rijn S, Meijer DH, Gainche L, Sena-Esteves M, Curry WT Jr, Carter BS, Krichevsky AM, Breakefield XO (2008) Glioblastoma microvesicles transport RNA and proteins that promote tumour growth and provide diagnostic biomarkers. Nat Cell Biol 10:1470–1476
2. Colombo M, Raposo G, Théry C (2014) Biogenesis, secretion, and intercellular interactions of exosomes and other extracellular vesicles. Ann Rev Cell Dev Biol 30:255–289
3. Gercel-Taylor C, Atay S, Tullis RH, Kesimer M, Taylor DD (2012) Nanoparticle analysis of

circulating cell-derived vesicles in ovarian cancer patients. Anal Biochem 428:44–53

4. Sokolova V, Ludwig AK, Hornung S, Rotan O, Horn PA, Epple M, Giebel B (2011) Characterisation of exosomes derived from human cells by nanoparticle tracking analysis and scanning electron microscopy. Colloids Surf B Biointerfaces 87:146–150
5. Dragovic RA, Gardiner C, Brooks AS, Tannetta DS, Ferguson DJ, Hole P, Carr B, Redman CW, Harris AL, Dobson PJ, Harrison P, Sargent IL (2011) Sizing and phenotyping of cellular vesicles using nanoparticle tracking analysis. Nanomedicine 7(6):780–788
6. Oosthuyzen W, Sime NE, Ivy JR, Turtle EJ, Street JM, Pound J, Bath LE, Webb DJ, Gregory CD, Bailey MA, Dear JW (2013) Quantification of human urinary exosomes by nanoparticle tracking analysis. J Physiol 591(23):5833–5842
7. Soo CY, Song Y, Zheng Y, Campbell EC, Riches AC, Gunn-Moore F, Powis SJ (2012) Nanoparticle tracking analysis monitors microvesicle and exosome secretion from immune cells. Immunology 136(2):192–197
8. King HW, Michael MZ, Gleadle JM (2012) Hypoxic enhancement of exosome release by breast cancer cells. BMC Cancer 12:421
9. Medintz IL, Uyeda HT, Goldman ER, Mattoussi H (2005) Quantum dot bioconjugates for imaging, labelling and sensing [review]. Nat Mater 4(6):435–446
10. Escola JM, Kleijmeer MJ, Stoorvogel W, Griffith JM, Yoshie O, Geuze HJ (1998) Selective enrichment of tetraspan proteins on the internal vesicles of multivesicular endosomes and on exosomes secreted by human B-lymphocytes. J Biol Chem 273:20121–20127
11. Heijnen HFG, Schiel AE, Fijnheer R, Geuze HJ, Sixma JJ (1999) Activated platelets release two types of membrane vesicles: microvesicles by surface shedding and exosomes derived from exocytosis of multivesicular bodies and alpha-granules. Blood 94:3791–3799
12. Thery C, Amigorena S, Raposo G, Clayton A (2006) Isolation and characterization of exosomes from cell culture supernatants and biological fluids. Curr Protoc Cell Biol 3:22
13. Pisitkun T, Shen R-F, Knepper MA (2004) Identification and proteomic profiling of exosomes in human urine. Proc Natl Acad Sci U S A 101(36):13368–13373
14. Gardiner C, Ferreira YJ, Dragovic RA, Redman CW, Sargent IL (2013) Extracellular vesicle sizing and enumeration by nanoparticle tracking analysis. J Extracell Vesicles 2:19671

Chapter 4

Imaging and Quantification of Extracellular Vesicles by Transmission Electron Microscopy

Romain Linares, Sisareuth Tan, Céline Gounou, and Alain R. Brisson

Abstract

Extracellular vesicles (EVs) are cell-derived vesicles that are present in blood and other body fluids. EVs raise major interest for their diverse physiopathological roles and their potential biomedical applications. However, the characterization and quantification of EVs constitute major challenges, mainly due to their small size and the lack of methods adapted for their study. Electron microscopy has made significant contributions to the EV field since their initial discovery. Here, we describe the use of two transmission electron microscopy (TEM) techniques for imaging and quantifying EVs. Cryo-TEM combined with receptor-specific gold labeling is applied to reveal the morphology, size, and phenotype of EVs, while their enumeration is achieved after high-speed sedimentation on EM grids.

Key words Blood plasma, Extracellular vesicles (EVs), Transmission electron microscopy (TEM), Cryo-TEM, Immuno-gold labeling, Annexin-A5 (Anx5), Phosphatidylserine (PS), Antibodies (Abs), Erythrocytes, Platelets

1 Introduction

Extracellular vesicles (EVs) are cell fragments enclosed within a lipid membrane, which are released by cells under various stimuli and are found in blood and other body fluids [1–4]. EVs, also called microvesicles, microparticles, or exosomes, have been proposed to participate in diverse functions in health and disease, which explains the intense research activity focusing on EVs. In blood, EVs participate in physiological processes of coagulation, inflammation and intercellular communication [5–7], while elevated EV levels have been reported in various diseases [8–11]. Other studies have stressed the potential biomedical applications of EVs as disease biomarkers, drug delivery vehicles, or vaccines [12–15].

Our current understanding on EVs—what they are and what they do—remains however limited, mainly because EVs are small

Andrew F. Hill (ed.), *Exosomes and Microvesicles: Methods and Protocols*, Methods in Molecular Biology, vol. 1545, DOI 10.1007/978-1-4939-6728-5_4, © Springer Science+Business Media LLC 2017

objects, most of them being comprised between 100 and 500 nm diameter [16], which renders difficult their characterization and their isolation. In addition, EVs originate from different cell types, hence are heterogeneous in composition, and they are present only at low concentration in unprocessed body fluids [17]. The difficulty of characterizing EVs is well recognized and is reflected by diverging results from the literature, e.g., on the size or the concentration of EVs in blood.

Many methods have been applied to the characterization of the structure or phenotype of EVs, including flow cytometry, transmission electron microscopy (TEM), atomic force microscopy, dynamic light scattering, nanoparticle tracking analysis, and resistive pulse sensing [18–20]. EM techniques used in combination with immuno-gold labeling have made significant contribution in the EV field, from the initial demonstration of their existence more than 30 years ago [21–23] to the recent description of their detailed structure [16, 17]. EVs have been analyzed extensively by classical electron microscopy (EM) techniques like negative staining [24, 25], thin sectioning [19], cryo-sectioning [26], or scanning-EM [27]. EV imaging by cryo-TEM has recently become popular [16, 28, 29], allowing for the observation of EVs in their hydrated near-native state [30].

Using cryo-TEM together with receptor-specific gold labeling, we have been able to provide a detailed description of the morphology, size distribution, and phenotypes of the main EV populations present in platelet free plasma (PFP) from healthy donors [16]. This approach, which presents the major advantage to be applicable to complex media such as unprocessed plasma, presents an intrinsic limitation for quantifying EVs. Indeed, as EVs are drained through a perforated carbon film covering EM grids, EVs larger than the carbon film openings, like tubular EVs and large fragments, are preferentially retained by the carbon net, while spherical EVs, which are smaller, pass freely through the net. Consequently, the enumeration of EVs by this approach is affected by a systematic error, called "fish-net" artifact, causing an over-estimation of the proportion of large EVs (Fig. 1). This problem could be overcome by using a simple method in which EVs were sedimented on EM grids covered with a continuous carbon film and quantified by TEM after air-drying. This method, previously used for the enumeration of viruses [31], allowed us to determine EV concentrations in PFP samples (Fig. 2).

We present here step-by-step application protocols of these two complementary TEM methods in the case of blood plasma samples (*see* **Note 1**).

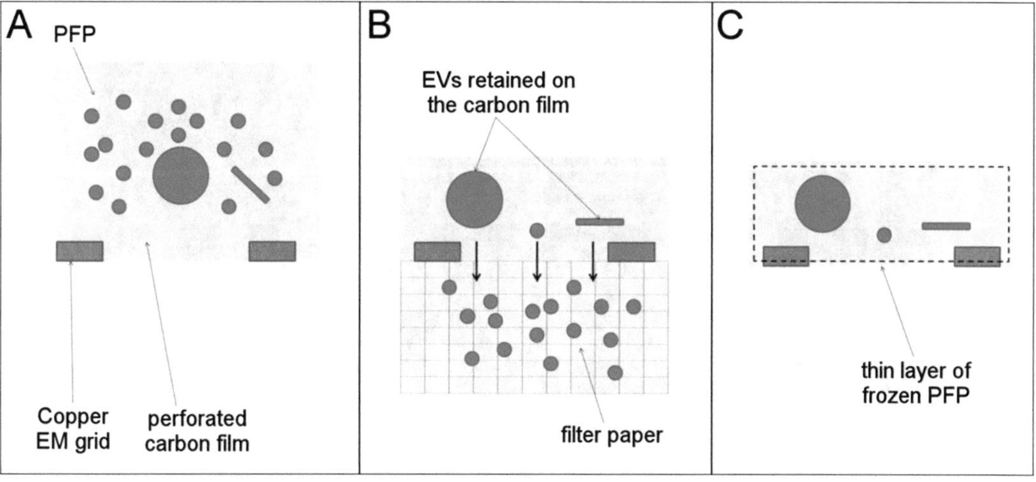

Fig. 1 Scheme of the "fish-net" artifact of cryo-TEM analysis of PFP samples. (a) PFP suspension layered over an EM grid coated with a perforated carbon net. The PFP contains EVs of various sizes. The scheme is not drawn to scale, as the distance between two copper bars is about 50 μm, the size of the holes in the carbon film varies from few hundred nm to several μm and the size of the EVs varies from 30 nm to 8 μm. (b) Upon draining the liquid with a filter paper from the bottom or back side of the EM grid, the large EVs are retained by the net, while the smaller EVs pass freely through the net. (c) Ultimately, a thin liquid layer of PFP of few hundred nm thickness is quickly frozen and observed in the microscope. Due to this "fish-net" artifact, the proportion of large EVs is over-estimated by cryo-TEM. Adapted from [16]

2 Materials

Prepare all solutions using ultrapure water (prepared by purifying deionized water to a resistivity of 18.2 MΩ cm with a tandem RiOs5-Synergy system (Millipore, Molsheim, France)) and analytical grade reagents (purchased from Sigma-Aldrich (Lyon, France) except when otherwise stated).

2.1 Blood Collection and PFP Preparation

1. 21-G needle.
2. 4.5 mL BD Vacutainer® tubes containing 0.1 volume of 129 mM sodium citrate (Becton Dickinson, Le Pont de Claix, France).
3. Eppendorf 5804-R centrifuge equipped with an A-4-44 swinging bucket rotor (Eppendorf, Montesson, France).

2.2 Sample Preparation

1. Phe-Pro-Arg-Chloromethyl ketone (PPACK) (Haematologic Technologies, Cryopep, Montpellier, France).
2. Gold nanoparticles conjugated [32] with:
   (a) Annexin-A5 (Anx5).
   (b) Anti-CD235a-mAb (clone 11E4B-7-6, Beckman Coulter, Villepinte, France).
   (c) Anti-CD41-mAb (clone P2, Beckman Coulter).

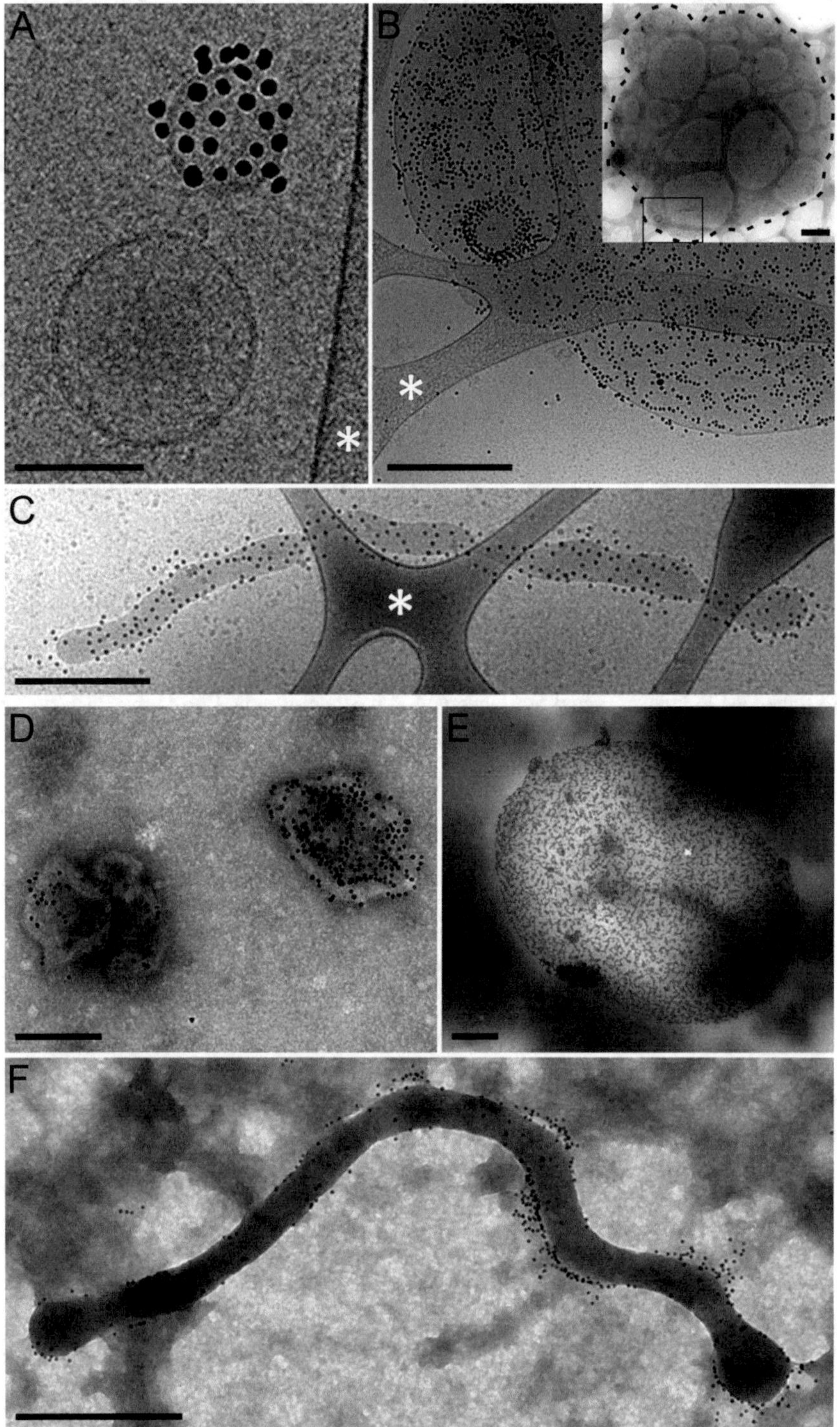

Fig. 2 Gallery of TEM images of EVs from PFP samples. (**a–c**), EVs imaged by Cryo-TEM after gold-labeling. (**a**) Spherical EVs after labeling with Anx5-gold-NPs, illustrating the high labeling specificity. Scale bar: 100 nm. (**b**) (*Inset*) Low magnification view of a large membrane fragment (contoured with a *back dashed line*). Scale bar: 1 μm. The image represents an enlarged view of the boxed area in the *inset*, showing the dense and

3. Cacodylate buffer (*see* **Note 2**): 100 mM sodium cacodylate, 0.02 % sodium azide, pH 7.4. Add about 900 mL water to a 1-L glass beaker. Weigh 21.4 g sodium cacodylate and transfer to the glass beaker. Add 2 mL of 10 % sodium azide and adjust pH to 7.4 with 1 M HCl. Make up to 1 L with water. Store at 4 °C.
4. Cacodylate buffer supplemented with Ca^{2+}: 100 mM sodium cacodylate, 2 mM $CaCl_2$, 0.02 % sodium azide, pH 7.4. Prepare as in **step 3**, except adding 400 μL of a 5 M $CaCl_2$ solution to the beaker before pH adjustment.
5. Polyallomer 4.5 mL centrifuge tubes (Beckman Coulter).
6. Standard razor blades.
7. Standard coarse (P80) and fine (P1200) sandpapers.
8. Home-made 12-mm diameter hemispherical piece of Epon-Araldite resin. Resin components (*see* **Note 3**) (Agar Scientific, Stansted, UK):
   (a) Agar 100 (exchangeable with Epon 812) and Araldite CY212 resins.
   (b) Hardener: dodecenylsuccinic anhydride (DDSA).
   (c) Accelerator: 2,4,6-tris dimethylaminomethyl phenol (DMP-30), exchangeable with benzyl dimethylamine (BDMA).
9. Drying oven.
10. Standard thin (less than 0.5 mm) double sided tape (Tesa, Lieusaint, France).
11. 300 mesh copper EM grids coated with a continuous carbon film (Ted Pella, Eloise, Tremblay-en-France, France).
12. Optima™ MAX-E ultracentrifuge equipped with a MLS50 rotor (Beckman Coulter).
13. 300 mesh copper EM grids coated with a perforated carbon film (Ted Pella).
14. UV/Ozone device (BHK, Claremont, CA, USA).
15. Leica EM-CPC cryo-chamber (Leica Microsystems SAS, Nanterre, France).
16. Whatman filter paper n°4/5 (Whatman, Versailles, France).
17. Cryo grid boxes (Eloise).
18. Liquid nitrogen cryo container (Dutscher, Brumath, France).
19. Gatan 626 cryo holder (Gatan, CA, USA).

Fig. 2 (continued) homogenous labeling with Anx5-gold-NPs. Scale bar: 0.5 μm. (**c**) Tubular EV labeled with anti-CD235a-gold-NPs. Scale bar: 0.5 μm. *White asterisks* in **a–c** point to areas of the carbon film. (**d–f**), Images of EVs observed on EM grids after sedimentation. (**d**) EVs presenting a near-circular shape, covered with Anx5-gold-NPs. Scale bar: 200 nm. (**e**) Large fragment homogenously labeled with Anx5-gold-NPs. Scale bar: 1 μm. (**f**) Tubular EV labeled with anti-CD235a-gold-NPs. Scale bar: 0.5 μm. In (**a–f**), very rare gold-NPs are present in the background, illustrating the high specificity of labeling. Adapted from [16] and [17]

3 Methods

Carry out all procedures at room temperature.

3.1 Fabrication of Resin Support for EM Grids (Fig. 3)

1. Mix 2.9 g Agar 100 resin, 2.35 g Araldite CY212 resin, 6.2 g DDSA and add 375 μL DMP-30.
2. Gently agitate the mix with a magnetic stirrer until homogeneity whilst avoiding the formation of bubbles (the mix is highly viscous).
3. Pour the mix with a syringe into 4.5 mL polyallomer centrifuge tubes, up to about 1 cm height.
4. Place the tubes in upright position in an oven at 60 °C for overnight polymerization.
5. Cut the tube wall with a razor blade and recover the solid resin pieces (*see* **Note 4**).
6. Sand their flat top surface successively with a coarse and a fine sandpaper (*see* **Note 5**).
7. If some of the resin mix is left over, it can be stored for several months at −20 °C and used later. Thaw frozen resin samples at room temperature, then follow **steps 3–6** in Subheading 3.1.

3.2 Blood Collection and Preparation of PFP Samples

1. Peripheral venous blood is drawn from donors after written informed consent. Blood is collected using a 21-G needle in 4.5 mL tubes containing 0.1 volume of 129 mM sodium citrate.
2. Discard the first tube, handle blood tubes with care and start the preparation of PFP within less than 1 h after blood collection [33].
3. Centrifuge blood tubes at 2500 ×*g* for 15 min at 25 °C in a centrifuge equipped with a swinging bucket rotor, with a low deceleration setting. Transfer the upper 1.8 mL of each tube in a 2-mL Eppendorf tube.
4. Centrifuge the 2-mL Eppendorf tubes at 2500 ×*g* for 15 min at 25 °C. Harvest and pool the upper 1.3 mL of each tube, which constitutes the PFP.

3.3 Preparation of PFP Samples for Cryo-TEM

3.3.1 Preparation of Pure PFP Samples

1. Expose EM grids coated with a perforated carbon film for 10 min to UV/ozone in order to render them hydrophilic [34] (*see* **Note 6**).
2. Deposit a 4 μL droplet of pure PFP onto an EM grid (*see* **Note 7**).
3. Remove the excess of liquid by blotting from the back side of the grid with a filter paper (*see* **Note 8**) and quickly plunge the grid into liquid ethane cooled down by liquid nitrogen (*see* **Note 9**) using a LEICA EM-CPC cryo-chamber.

4. Maintain EM grids prepared this way under liquid nitrogen and store them in cryo grid boxes in a cryo container until use (*see* **Note 10**).
5. For cryo-TEM observation, mount a grid in a cryo-holder and insert it in the electron microscope.

3.3.2 Preparation of PFP Samples Labeled with Gold Nanoparticles Conjugated with Proteins

1. Deposit 7 μL of fresh PFP in an Eppendorf tube, add 1 μL of 100 μM PPACK solution prepared in water and homogenize gently with a micropipette.
2. For labeling PS-exposing EVs, add 1 μL of Anx5-conjugated gold nanoparticles (Anx5-gold-NPs) at a concentration of 2×10^{16} particles/L, 1 μL of 100 mM $CaCl_2$ and homogenize gently with a micropipette (*see* **Note 11**).
3. Incubate for 15 min.
4. For labeling EVs by means of specific antibody-conjugated gold nanoparticles (Ab-gold-NPs), apply the same procedure, except that PPACK and $CaCl_2$ are omitted and substituted by 2 μL PFP and incubation time is 1 h (*see* **Note 12**).
5. For double labeling experiments, label first PFP samples with 10 nm Ab-gold-NPs for 45 min, then with 4 nm Anx5-gold-NPs for an additional 15 min.
6. Then, follow **steps 1–5** in Subheading 3.3.1.

3.4 Sedimentation of PFP Samples on EM Grids

1. For labeling PS-exposing EVs, deposit 100 μL of fresh PFP (*see* **Note 7**) in an Eppendorf tube, add 1 μL of 1 mM PPACK, 1 μL of Anx5-gold-NPs at 2×10^{16} particles/L, and 5 μL of 200 mM $CaCl_2$. Homogenize gently with a pipette.
2. Incubate for 15 min.
3. For labeling EVs with specific Ab-gold-NPs, apply the same procedure except that PPACK and $CaCl_2$ are omitted and incubation time is 1 h (*see* **Note 13**).
4. Take a 12-mm diameter resin support, adhere a thin piece of double-sided tape and fix 2–4 EM grids coated with a continuous carbon film. Only the grid edges should contact the tape in order to recover them easily (*see* Fig. 3).
5. Deposit the resin support with the EM grids at the bottom of a 4.5 mL polyallomer centrifuge tube.
6. Transfer the volume from the Eppendorf tube over the resin support.
7. Add gently 3 mL of 100 mM cacodylate buffer, pH 7.4, supplemented with 2 mM $CaCl_2$ in the case of Anx5 labeling.
8. Centrifuge in an ultracentrifuge equipped with a swinging-bucket rotor at $100{,}000 \times g$ for 1 h at 20 °C.
9. Remove carefully the liquid above the EM grids and discard it.

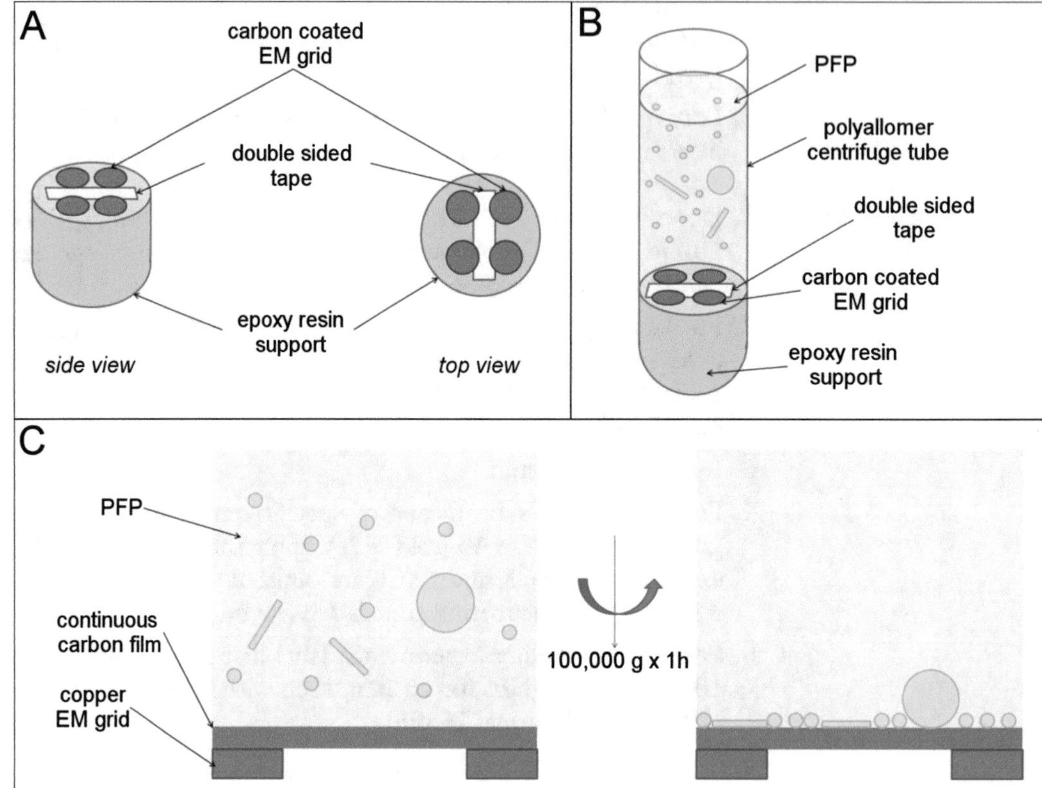

Fig. 3 Scheme of the procedure of EV sedimentation on EM grids. (**a**) Side and top views of an epoxy resin support with a hemispherical end on one side and a flat surface on the other side. A thin piece of double-sided tape is fixed on the flat surface and four EM grids coated with a continuous carbon film are deposited on the tape edges. (**b**) The epoxy support with four attached EM grids is inserted in a centrifuge tube before filling the tube with a PFP suspension. (**c**) Scheme of an EM grid coated with a continuous carbon film with an overlying PFP suspension before (*left*) and after (*right*) centrifugation. Adapted from [16]

10. Recover the EM grids by cutting the centrifuge tube with a razor blade, let them dry in air, and store them in an EM-grid box (*see* **Note 14**).

3.5 Transmission Electron Microscopy Observation and EV Quantification

TEM observation of PFP samples is carried out similarly for both cryo samples and samples sedimented on EM grids. However, distinct procedures are applied due to large differences in the overall number of EVs per grid square.

3.5.1 Cryo-TEM Observation

1. Scan the grid first at 300× magnification over several hundred squares (*see* **Note 15**). This step allows the detection and quantification of large (over 1 μm) EVs.
2. Second, scan the grid at 5000× magnification (*see* **Note 16**). This is required for the detection of sub-micrometer EVs.

3.5.2 TEM Observation After On-Grid Sedimentation and EV Quantification

1. Scan the grid first at 300× magnification over several hundred squares (*see* **Note 15**). This step allows the detection and quantification of large (over 1 μm) EVs.
2. Second, scan fives grid squares (*see* **Note 17**) at 5000× magnification (*see* **Note 18**). This is required for the detection of sub-micrometer EVs. As this step is more time-consuming, a manual cell-counter is handy and limits operator-induced bias in EV quantification.
3. Apply a stringent method for enumerating EVs, in which only the objects that present a high and homogenous gold-NP labeling with a well-defined near-to-circular or tubular shape are taken into account (*see* Fig. 2).
4. Calculate the EV concentration in pure PFP, by taking into account the number (N) of EVs counted per μm^2, the total area of sedimentation (A in μm^2) the nominal volume (V in μL) of PFP used for centrifugation. The EV concentration per μL pure PFP is equal to $(N \times A)/V$.

4 Notes

1. The methods described here for PFP samples are in principle applicable with other body fluids, as already established for cryo-TEM with synovial liquid [35], milk [36], pleural liquid (Ms submitted) or cerebrospinal fluid [37], as well as for cell culture EVs. Each type of material requires specific conditions of preparation, principally for eliminating cells, cell debris, and large aggregates contained in the original sample.
2. Cacodylate buffer contains arsenic and sodium azide, which are toxic compounds. Wear recommended protection equipments (gloves and mask for weighting and gloves for use).
3. Wear gloves for weighting and using resin components which can be harmful or corrosive.
4. Be careful when using razor blades.
5. Wear a mask when sanding resin supports, to avoid inhaling resin dust.
6. Glow discharge devices can also be used.
7. Homogenize PFP before use by gentle pipetting.
8. This step requires some practice. The duration of the blotting step is critical and depends on the viscosity of the liquid medium.
9. Wear gloves, protection glasses and long-sleeved clothes, as liquid ethane may cause severe burnings.
10. Maintenance of the cold chain is mandatory, as vitreous ice only exists at T below −130 °C. A temperature around −170 °C is ideal.

11. A final volume of 10 μL allows preparing two cryo-EM grids per sample, which is highly recommended because the success rate in the preparation of cryo-EM grids of good quality is variable.
12. Incubation time must be optimized with each other type of Ab-conjugated gold-NPs.
13. Although double labeling can in principle be performed with the on-grid sedimentation approach by using gold-NPs of two distinct sizes, e.g., 10 and 4 nm as described in cryo-TEM in Subheading 3.3.2.4, the detection of small (<10 nm) gold-NPs requires the use of a higher magnification (e.g., 13,000 × *g* for 4 nm particles), which makes the scanning operation much longer.
14. We noticed a variable aging of EM grids prepared this way, probably due to environmental storage conditions, principally humidity level. We recommend observing shortly after preparation, ideally within a week.
15. With PFP from healthy donors, about one to five large EVs are observed over a total of one hundred grid squares.
16. Absolute quantification cannot be achieved by cryo-TEM due to the "fish-net" artifact described in Fig. 1 and also because the overall EV concentration in PFP is low, of the order of 100,000 EVs per μL [17]. On average, less than one EV per grid square is observed with PFP from healthy donors. Increasing EV concentration for example by ultracentrifugation is possible [25]. However this may lead to modifications of the original material, such as vesicle aggregation, fragmentation of large objects, or activation of residual platelets.
17. As the repartition of EVs over the grid may be heterogeneous, it is recommended to select squares at different places over the grid, in order to obtain representative data.
18. With PFP from healthy donors, about one hundred EVs are observed per grid square with the on-grid sedimentation method.

Acknowledgment

This study was supported by ANR (grant 11-BSV1-03501-PlacentA5 to A.B.).

References

1. Théry C, Ostrowski M, Segura E (2009) Membrane vesicles as conveyors of immune responses. Nat Rev Immunol 9:581–593
2. György B, Szabó TG, Pásztói M, Pál Z, Misják P, Aradi B, László V, Pállinger E, Pap E, Kittel A, Nagy G, Falus A, Buzás EI (2011) Membrane vesicles, current state-of-the-art: emerging role of extracellular vesicles. Cell Mol Life Sci CMLS 68:2667–2688
3. Van der Pol E, Böing AN, Harrison P, Sturk A, Nieuwland R (2012) Classification, functions, and clinical relevance of extracellular vesicles. Pharmacol Rev 64:676–705
4. Raposo G, Stoorvogel W (2013) Extracellular vesicles: exosomes, microvesicles, and friends. J Cell Biol 200:373–383
5. Zwicker JI, Trenor CC, Furie BC, Furie B (2011) Tissue factor-bearing microparticles and thrombus formation. Arterioscler Thromb Vasc Biol 31:728–733
6. Morel O, Toti F, Hugel B, Bakouboula B, Camoin-Jau L, Dignat-George F, Freyssinet J-M (2006) Procoagulant microparticles: disrupting the vascular homeostasis equation? Arterioscler Thromb Vasc Biol 26:2594–2604
7. Valadi H, Ekström K, Bossios A, Sjöstrand M, Lee JJ, Lötvall JO (2007) Exosome-mediated transfer of mRNAs and microRNAs is a novel mechanism of genetic exchange between cells. Nat Cell Biol 9:654–659
8. VanWijk MJ, VanBavel E, Sturk A, Nieuwland R (2003) Microparticles in cardiovascular diseases. Cardiovasc Res 59:277–287
9. Boilard E, Nigrovic PA, Larabee K, Watts GFM, Coblyn JS, Weinblatt ME, Massarotti EM, Remold-O'Donnell E, Farndale RW, Ware J, Lee DM (2010) Platelets amplify inflammation in arthritis via collagen-dependent microparticle production. Science 327:580–583
10. Delabranche X, Boisramé-Helms J, Asfar P, Berger A, Mootien Y, Lavigne T, Grunebaum L, Lanza F, Gachet C, Freyssinet J-M, Toti F, Meziani F (2013) Microparticles are new biomarkers of septic shock-induced disseminated intravascular coagulopathy. Intensive Care Med 39:1695–1703
11. Manly DA, Wang J, Glover SL, Kasthuri R, Liebman HA, Key NS, Mackman N (2010) Increased microparticle tissue factor activity in cancer patients with venous thromboembolism. Thromb Res 125:511–512
12. Skog J, Würdinger T, van Rijn S, Meijer DH, Gainche L, Sena-Esteves M, Curry WT, Carter BS, Krichevsky AM, Breakefield XO (2008) Glioblastoma microvesicles transport RNA and proteins that promote tumour growth and provide diagnostic biomarkers. Nat Cell Biol 10:1470–1476
13. Burger D, Schock S, Thompson CS, Montezano AC, Hakim AM (1979) Touyz RM (2013) Microparticles: biomarkers and beyond. Clin Sci Lond Engl 124:423–441
14. Alvarez-Erviti L, Seow Y, Yin H, Betts C, Lakhal S, Wood MJA (2011) Delivery of siRNA to the mouse brain by systemic injection of targeted exosomes. Nat Biotechnol 29:341–345
15. Zitvogel L, Regnault A, Lozier A, Wolfers J, Flament C, Tenza D, Ricciardi-Castagnoli P, Raposo G, Amigorena S (1998) Eradication of established murine tumors using a novel cell-free vaccine: dendritic cell derived exosomes. Nat Med 4:594–600
16. Arraud N, Linares R, Tan S, Gounou C, Pasquet J-M, Mornet S, Brisson AR (2014) Extracellular vesicles from blood plasma: determination of their morphology, size, phenotype and concentration. J Thromb Haemost 12:614–627
17. Arraud N, Gounou C, Linares R, Brisson AR (2015) A simple flow cytometry method improves the detection of phosphatidylserine-exposing extracellular vesicles. J Thromb Haemost 13(2):237–247. doi:10.1111/jth.12767
18. Van der Pol E, Hoekstra AG, Sturk A, Otto C, van Leeuwen TG, Nieuwland R (2010) Optical and non-optical methods for detection and characterization of microparticles and exosomes. J Thromb Haemost 8:2596–2607
19. György B, Szabó TG, Turiák L, Wright M, Herczeg P, Lédeczi Z, Kittel Á, Polgár A, Tóth K, Dérfalvi B, Zelenák G, Böröcz I, Carr B, Nagy G, Vékey K, Gay S, Falus A, Buzás EI (2012) Improved flow cytometric assessment reveals distinct microvesicle (cell-derived microparticle) signatures in joint diseases. PLoS ONE 7:e49726
20. Witwer KW, Buzás EI, Bemis LT, Bora A, Lässer C, Lötvall J, Nolte-'t Hoen EN, Piper MG, Sivaraman S, Skog J, Théry C, Wauben MH, Hochberg F (2013) Standardization of sample collection, isolation and analysis meth-

ods in extracellular vesicle research. J Extracell Vesicles. doi:10.3402/jev.v2i0.20360

21. Wolf P (1967) The nature and significance of platelet products in human plasma. Br J Haematol 13:269–288
22. Harding C, Heuser J, Stahl P (1983) Receptor-mediated endocytosis of transferrin and recycling of the transferrin receptor in rat reticulocytes. J Cell Biol 97:329–339
23. Pan BT, Teng K, Wu C, Adam M, Johnstone RM (1985) Electron microscopic evidence for externalization of the transferrin receptor in vesicular form in sheep reticulocytes. J Cell Biol 101:942–948
24. Heijnen HF, Schiel AE, Fijnheer R, Geuze HJ, Sixma JJ (1999) Activated platelets release two types of membrane vesicles: microvesicles by surface shedding and exosomes derived from exocytosis of multivesicular bodies and alpha-granules. Blood 94:3791–3799
25. Böing AN, van der Pol E, Grootemaat AE, Coumans FAW, Sturk A, Nieuwland R (2014) Single-step isolation of extracellular vesicles by size-exclusion chromatography. J Extracell Vesicles 3:PMCID: PMC4159761. doi:10.3402/jev.v3.23430
26. Théry C, Amigorena S, Raposo G, Clayton A (2006) Isolation and characterization of exosomes from cell culture supernatants and biological fluids. Curr Protoc Cell Biol Editor Board Juan Bonifacino Al Chapter 3:Unit 3.22.
27. Kralj-Iglic V, Sustar B-Z, Stukelj F, Bobojevic J, Ogorevc K, Mam S, Mancek-Keber J, Rozman VP, Hagerstrand K-IV (2011) Nanoparticles isolated from blood: a reflection of vesiculability of blood cells during the isolation process. Int J Nanomedicine 6:2737–2748
28. Tatischeff I, Larquet E, Falcón-Pérez JM, Turpin P-Y, Kruglik SG (2012) Fast characterisation of cell-derived extracellular vesicles by nanoparticles tracking analysis, cryo-electron microscopy, and Raman tweezers microspectroscopy. J Extracell Vesicles. doi:10.3402/jev.v1i0.19179
29. Yuana Y, Koning RI, Kuil ME, Rensen PCN, Koster AJ, Bertina RM, Osanto S (2013) Cryo-electron microscopy of extracellular vesicles in fresh plasma. J Extracell Vesicles. doi:10.3402/jev.v2i0.21494
30. Dubochet J, Adrian M, Chang JJ, Homo JC, Lepault J, McDowall AW, Schultz P (1988) Cryo-electron microscopy of vitrified specimens. Q Rev Biophys 21:129–228
31. Miller MF (1974) Principles and techniques of electron microscopy. Biol Appl 1974:89–128
32. Brisson AR, Mornet S (2007) Patent WO/2007/122259.
33. Lacroix R, Judicone C, Poncelet P, Robert S, Arnaud L, Sampol J, Dignat-George F (2012) Impact of pre-analytical parameters on the measurement of circulating microparticles: towards standardization of protocol. J Thromb Haemost 10:437–446
34. Richter R, Mukhopadhyay A, Brisson A (2003) Pathways of lipid vesicle deposition on solid surfaces: a combined QCM-D and AFM study. Biophys J 85:3035–3047
35. Cloutier N, Tan S, Boudreau LH, Cramb C, Subbaiah R, Lahey L, Albert A, Shnayder R, Gobezie R, Nigrovic PA, Farndale RW, Robinson WH, Brisson A, Lee DM, Boilard E (2013) The exposure of autoantigens by microparticles underlies the formation of potent inflammatory components: the microparticle-associated immune complexes. EMBO Mol Med 5:235–249
36. Zonneveld MI, Brisson AR, van Herwijnen MJC, Tan S, van de Lest CHA, Redegeld FA, Garssen J, Wauben MHM, Nolte-'t Hoen ENM (2014) Recovery of extracellular vesicles from human breast milk is influenced by sample collection and vesicle isolation procedures. J Extracell Vesicles. doi:10.3402/jev.v3.24215
37. Chen WW, Balaj L, Liau LM, Samuels ML, Kotsopoulos SK, Maguire CA, LoGuidice L, Soto H, Garrett M, Zhu LD, Sivaraman S, Chen C, Wong ET, Carter BS, Hochberg FH, Breakefield XO, Skog J (2013) BEAMing and droplet digital pcr analysis of mutant IDH1 mRNA in glioma patient serum and cerebrospinal fluid extracellular vesicles. Mol Ther Nucleic Acids 2:e109

Chapter 5

Quantitative Analysis of Exosomal miRNA via qPCR and Digital PCR

Shayne A. Bellingham, Mitch Shambrook, and Andrew F. Hill

Abstract

Extracellular vesicles, such as exosomes and microvesicles, have been shown to contain potential microRNA (miRNA) biomarkers that may be utilized in the diagnosis of various diseases from cancer to neurological disorders. The unique nature of the extracellular vesicle bilayer allows miRNA to be protected from degradation making it an ideal source of material for biomarkers discovery from both fresh and archived samples. Here we describe the quantitative analysis of miRNA isolated from exosomes by quantitative PCR and digital PCR.

Key words Exosomes, miRNA, Biomarker, Taqman assay, RNA isolation, Quantitative PCR, Digital PCR

1 Introduction

MicroRNAs (miRNAs) are a class of noncoding RNA species of ~22–24 nucleotides in length that functionally repress target mRNA by binding their 3′-untranslated regions [1]. Currently there are more than 2500 known miRNA species in the latest miRBASE release v21, with each potentially having multiple messenger RNA gene targets [2]. Subsequently, miRNA can regulate essential biological pathways such as cellular signaling, cellular differentiation and development, proliferation, infection and immunity response, tumorigenesis, and apoptosis [3].

Exosomes are small vesicles that are secreted into the extracellular milieu from a variety of cells and can be isolated from both cell culture media and body fluids such as blood, urine, saliva, cerebrospinal fluid and synovial fluid [4–9]. Exosomes contain both mRNA and miRNA, termed exosomal RNA, which has been shown to be functional when transferred between different cell types [6]. Exosomal RNA has also been demonstrated to be protected from degradation by cellular RNAses and relatively stable when stored for extended periods of time or subjected to temperature changes [4, 10].

Andrew F. Hill (ed.), *Exosomes and Microvesicles: Methods and Protocols,* Methods in Molecular Biology, vol. 1545, DOI 10.1007/978-1-4939-6728-5_5, © Springer Science+Business Media LLC 2017

Aberrant miRNA expression has been implicated in a number of diseases including cancer, cardiovascular disease and neurological disorders [11–13]. The role of exosomal miRNA as a disease biomarker has also been of great interest for the diagnosis of several diseases. Consequently, several exosomal miRNA signatures have been identified from circulating biological fluids and cell culture studies that show potential both as a biomarker for diseases such as cancer, neurodegenerative disorders, and kidney disease [9, 14–18].

In this chapter, we describe detection of exosomal miRNA signatures via quantitative PCR (qRT-PCR) and QuantStudio™ 3D Digital PCR System using TaqMan microRNA assays. However, other qRT-PCR and Digital PCR platforms are available, including Qiagen's miScript system, Exiqon's miRCURY LNA system, and Bio-Rad's Droplet Digital PCR System. The principal steps involved in analysis of miRNA signature require extraction of high-quality RNA that includes small RNA <200 nt in length and validation of specific miRNA. The techniques described here are for exosomes samples captured from cell culture medium via standard ultracentrifugation techniques or Optiprep Gradient fractionation [9, 19]. They can also be adapted for exosomal miRNA analysis from biological fluids such as urine, plasma, or serum samples [14, 20].

2 Materials

Prepare all solutions using 0.2 μM filtered ultrapure Milli-Q RNase free water and analytical grade reagents. Prepare and store all reagents at room temperature (unless indicated otherwise). It is essential that handling and collection procedures for biological samples are carried out according to strict protocols to ensure RNA does not degrade.

2.1 Extraction of Exosomal RNA

1. Microcentrifuge with ability to be set at 4 °C or placed in cold room.
2. Phosphate buffered saline (PBS).
3. miRNeasy Isolation Kit (Qiagen).
4. Trizol LS Reagent (Invitrogen).
5. Chloroform.
6. 100% ethanol, ACS grade.
7. Milli-Q RNase/DNase-free water.
8. Luer Lock syringes.
9. 0.2 μM syringe filters.
10. Heat block set to 95–100 °C.
11. Screw-capped RNase-free 1.5 mL Eppendorf tubes.

2.2 Exosomal MicroRNA Expression Analysis Using qRT-PCR

1. Agilent 2100 Bioanalyzer (Agilent Technologies Inc.).
2. Agilent RNA 6000 Nano Kit (Agilent Technologies Inc.).
3. Agilent Small RNA Kit (Agilent Technologies Inc.).
4. Taqman MicroRNA Assays for each specific miRNA (Applied Biosystems).
5. MicroRNA Reverse Transcription Kit.
6. Taqman Gene Expression Master Mix or Fast Advanced Master Mix (Applied Biosystems).
7. Viia7 Real-Time PCR system (Applied Biosystems), or similar instrument.
8. Fast-reaction tubes (0.1 mL), 96-well plates (Applied Biosystems), or equivalent.
9. microAmp Adhesive Film (Applied Biosystems).
10. GeneAmp® PCR System 9700 (Applied Biosystems).

2.3 Digital PCR Using QuantStudio 3D System

1. QuantStudio™ 3D Digital PCR 20K Chips, Chip Case Lids, and Sample Loading Blades (Applied Biosystems).
2. QuantStudio™ 12K Flex OpenArray® Immersion Fluid and Fluid Tip (Applied Biosystems).
3. QuantStudio™ 3D Digital PCR Chip Loader (Applied Biosystems).
4. UV-Activated Chip Sealant Syringe (Applied Biosystems).

2.4 Optional Materials

1. QIAcube (Qiagen), fully automated sample processing for Subheading 3.1.
2. TaqMan PreAmp Master Mix (Applied Biosystems), in Subheading 3.2.
3. QIAgility (Qiagen), for liquid handling and reaction plate setup in Subheading 3.2.

3 Methods

3.1 Extraction of Total RNA Including miRNA with miRNeasy Mini Kit from Exosome Samples

The miRNeasy Mini Kit enables purification of total RNA, which includes RNA from approximately >18 nucleotides from all types of animal tissues and cells, including difficult-to-lyse tissues. The miRNeasy Mini Kit is used for low-throughput RNA purification using spin columns. Co-purification of miRNA and total RNA using the miRNeasy Mini Kit can be automated on the QIAcube (Optional).

1. Cool centrifuge to 4 °C.
2. Ensure exosome samples are in ~250 μL 0.2 μM filtered PBS per extraction in screw-cap Eppendorf tubes (*see* **Note 1**).

3. Disrupt exosome sample in 750 μL of TRIzol LS reagent. Vortex the tube for 10 s (*see* **Note 2**).
4. Incubate the tube containing the homogenate at room temperature (15–25 °C) for 5 min.
5. Add 250 μL chloroform to the tube containing the homogenate in the fume hood and cap the tube securely. Vortex the tube for 15 s (*see* **Note 3**).
6. Incubate the tube containing the homogenate at room temperature for 2–3 min in the fume hood.
7. Centrifuge for 15 min at 12,000 × *g* at 4 °C (*see* **Note 4**).
8. Carefully transfer the upper aqueous phase to a new collection tube and record the volume. The volume of the aqueous phase should be recorded accurately. Do not transfer or disrupt the interphase or lower red organic phase as it contains protein and DNA contamination (*see* **Note 5**) (Optional QIAcube extraction *see* **Note 6**).
9. Add 1.5 volumes of 100% ethanol to the volume of "recovered aqueous phase from **step 8**" and mix thoroughly by inverting or pipetting several times. For example if 350 μL aqueous phase recovered add 525 μL of 100% ethanol. Do not centrifuge.
10. For each sample, assemble a miRNeasy Mini spin column together with a 2 mL collection tube (supplied in the miRNeasy kit).
11. Pipet up to 700 μL of the sample, including any precipitate that may have formed, into the assembled miRNeasy Mini spin column and centrifuge at 8000 × *g* (10,000 rpm) for 15–30 s at room temperature (15–25 °C). Discard the flow-through and reuse the collection tube.
12. Repeat **step 10** using the remainder of the sample. Discard the flow-through and reuse the collection tube.
13. Add 700 μL Buffer RWT to the miRNeasy Mini spin column. Centrifuge for 15–30 s at 8000 × *g* (10,000 rpm) to wash the column. Discard the flow-through, reusing the collection tube.
14. Pipet 500 μL Buffer RPE into the miRNeasy Mini spin column. Centrifuge for 15–30 s at 8000 × *g* (10,000 rpm) to wash the column. Discard the flow-through and reuse the collection tube.
15. Add another 500 μL Buffer RPE to the miRNeasy Mini spin column. Close the lid gently and centrifuge for 2 min at 8000 × *g* (10,000 rpm) to dry the miRNeasy Mini spin column membrane (*see* **Note 7**).

16. Place the miRNeasy Mini spin column into a new 2 mL collection tube (not supplied), and discard the old collection tube with the flow-through. Centrifuge in a microcentrifuge at full speed for 1 min (*see* **Note 8**).
17. Label clean RNase free 1.5 mL collection tubes and transfer the miRNeasy Mini spin column to each collection tube.
18. Pipet 30–50 μL nuclease-free water directly onto the miRNeasy Mini spin column membrane.
19. Centrifuge for 1 min at 8000 × *g* (10,000 rpm) to elute the RNA (*see* **Note 9**).
20. Eluted RNA should be immediately placed on ice to prevent potential degradation. Keep at −20 or −80 °C for short-and long-term storage, respectively.

3.2 Exosomal MicroRNA Expression Analysis Using qRT-PCR

A number of factors govern the choice of miRNAs chosen for qRT-PCR validation. Importantly, if cells and/or tissues are of murine origin, the candidate miRNA must have a known human homologue, so that putative disease biomarkers identified can be tested in human patient samples. Second, given the goal of producing clinically relevant assays, miRNAs should be highly detectable in both relevant diseased tissue and noninvasive body fluids such blood. Finally, prioritization of miRNAs based on their function, such as a known role in the target disease of interest, may be a sensible approach if the candidate list is large.

3.2.1 MicroRNA Reverse Transcription

The MicroRNA Reverse Transcription (RT) Kit provides the necessary components for optimal performance in TaqMan MicroRNA Assays. Components of this kit are used with the RT primer provided with the TaqMan MicroRNA Assay to convert miRNA to cDNA (*see* **Note 10**).

1. Assess the quantity and quality of isolated total RNA including microRNA by running 1 μL of the Total RNA sample on the Agilent Bioanalyzer using both the Agilent RNA 6000 kit and Agilent small RNA assay kit, following manufacturer's instructions (*see* **Note 11**).
2. Before use, thaw the 5× RT primer and MicroRNA Reverse Transcription Kit components on ice, resuspend the solutions completely by gently vortexing, then briefly centrifuge 1000 × *g*, ~10 s (*see* **Notes 12** and **13**).
3. In a polypropylene tube, prepare the RT master mix on ice by multiplying the volume of components in Table 1 to the required number of RT reactions (*see* **Note 14**).
4. Mix gently. *Do not vortex.* Centrifuge in a microcentrifuge to bring the solution to the bottom of the tube (*see* **Note 13**).
5. Place the RT master mix on ice until you prepare the RT reaction tubes or plate.

Table 1
Master Mix for reverse transcription reactions

Component	RT master mix volume per 15 μL reaction	Volume for ten reactions including excess
10× RT buffer	1.5	18.0
25× dNTP mix (100 mM)	0.15	1.8
RNase inhibitor	0.19	2.28
MultiScribe™ reverse transcriptase	1.0	12.0
Nuclease-free milli Q H_2O	4.16	49.92
Total per reaction	7.0	84.0

Table 2
Thermal cycler conditions for reverse transcription

	Step 1	Step 2	Step 3	Step 4
Temperature (°C)	16	42	85	4
Time	30 min	30 min	5 s	∞

6. For each 15-μL RT reaction, combine into a 0.2-mL polypropylene RT reaction tube or into a well of a 96-well reaction plate 7 μL RT master mix (from step 3) with 5 μL total RNA (1–10 ng per reaction as determined from concentration provided by Bioanalyzer in Subheading 3.3.1, **step 1**) (*see* **Note 15**).
7. Add 3 μL of 5× RT primer from each assay set into the corresponding RT reaction tube or plate well.
8. Seal the tube with optical caps or reaction plate with optical adhesive film and mix thoroughly by inverting the solution. Centrifuge to bring the solution to the bottom of the tube or well.
9. Incubate the tube on ice for 5 min and keep it on ice until you are ready to load the thermal cycler.
10. To perform reverse transcription: Program the thermal cycler conditions as in Table 2.
11. Load the reactions into the thermal cycler and start the reverse transcription. When run has finished, remove tubes and centrifuge briefly to spin down contents and eliminate any air bubbles. Store cDNA for short-term (4 °C) or long-term storage (−15 to −25 °C) or proceed immediately to qPCR.

3.2.2 Taqman Gene Expression MicroRNA Assay

The TaqMan® MicroRNA Assays combine a target-specific stem-loop reverse transcription primer that extends the 3′ end of the target to produce a cDNA template that allows specific detection of mature miRNA with a microRNA gene expression assay that can be used in standard real-time PCR.

1. If frozen, thaw cDNA samples by placing them on ice. When thawed, resuspend the samples by vortexing and then centrifuge the tubes briefly.
2. Thaw 20× Taqman microRNA assays on ice, and resuspend samples by vortexing, then centrifuge the tubes at 1000 × *g*, for 10 s (*see* **Notes 12** and **16**).
3. Thaw Taqman Gene Expression Master mix or Fast Advanced Master mix on ice and mix by thoroughly by swirling the bottle.
4. Add 210 μL nuclease-free milli Q H_2O to each individual cDNA template for a final dilution of 1:15, mix and centrifuge briefly (*see* **Note 17**).
5. Prepare the PCR reaction mix in a microcentrifuge tube on ice according to Table 3 before transferring it to the reaction plate for thermal cycling and fluorescence analysis (*see* **Note 18**).
6. Vortex then centrifuge the tube(s) briefly to spin down the contents and eliminate any air bubbles from the solutions.
7. Transfer the appropriate volume of each reaction mixture to each well of the Applied Biosystems 96-well plate, seal plate with optical adhesive film and spin for 5 min, 2000 × *g* at 4 °C in a centrifuge fitted with plate adapters (*see* **Note 19**).

Table 3
Reaction mixtures for TaqMan gene expression assays

	20 μL reactions		10 μL reactions	
	1 replicate	**4 replicates**	**1 replicates**	**4 replicates**
TaqMan gene expression (2×) or fast advanced master mix (2×)	10.0	50.0	5.0	25.0
TaqMan microRNA assay (20×)	1.0	5.0	0.5	2.5
Diluted cDNA template	9.0	45.0	4.5	22.5
Total volume	20.0	100.0	10.0	50.0

Table 4
qPCR thermal cycling conditions

Standard mode thermal-cycling conditions				
	UDG incubation	Polymerase activation	PCR (40 cycles)	
Step	Hold	Hold	Denature	Anneal/extend
Time	2 min	10 min	15 s	1 min
Temp.	50 °C	95 °C	95 °C	60 °C

Fast mode thermal-cycling conditions				
	UNG incubation	Polymerase activation	PCR (40 cycles)	
Step	Hold	Hold	Denature	Anneal/extend
Time	2 min	20 s	1 s	20 s
Temp	50 °C	95 °C	95 °C	60 °C

8. Load and run the plate on real-time quantitative PCR instrument, Viia7 or equivalent instrument either standard mode or fast mode as indicated in Table 4 (*see* **Note 20**).
9. The relative fold change of miRNA levels are determined based on the threshold cycle (Ct) and are normalized by the ΔΔCt method [21]. Normalize your Ct values using the average CT of the endogenous controls (*see* **Notes 21** and **22**).
10. Use ΔCT (miRNA Ct—averaged endogenous control Ct) or fold-change relative to a calibrator or reference sample ($2^{\Delta\Delta CT}$) for relative expression analysis.

3.3 Quantstudio™ 3D Digital PCR Assay

This PCR assay utilizes chips containing 20,000 nanoscale reaction wells partitioning a sample into as many as 20,000 independent PCR reactions. This enables higher sensitivity, accuracy and determination of an absolute concentration of target copies/μL without a standard curve or reference.

3.3.1 Prepare the Reaction Mix and Samples

1. Remove from storage and allow the QuantStudio™ 3D Digital PCR Master Mix and TaqMan® Assay(s) to reach room temperature.
2. Review the concentration of your cDNA samples and prepare dilutions if necessary (*see* **Note 23**).
3. Gently invert the tube of Digital PCR Master Mix ten times (or gently vortex on low–medium speed).

4. In an 0.5- or 1.5-mL reaction tube, prepare sufficient PCR reaction mix for your samples. Scale the component amounts appropriately, depending on the number of samples that you are running.
5. Vortex, then briefly centrifuge the DNA samples.
6. Transfer 32.4 μL of PCR reaction mix to each labeled reaction tube.
7. Transfer 3.6 μL of each sample, diluted to the appropriate concentration, to the corresponding reaction tube. Mix well by gently pipetting up and down after each transfer (or gently vortex on low-medium speed).
8. Cap the reaction tubes, then briefly centrifuge them and immediately proceed to load the Digital PCR 20K Chips (*see* **Note 24**).

3.3.2 Load the Chips Using the Chip Loader

1. Allow the prepared digital PCR reaction to equilibrate to room temperature (approximately 15 min) and remove the following consumables from their packaging and place them on a clean, dry, lint-free surface:
   - QuantStudio™ 3D Digital PCR Chip Case Lid.
   - QuantStudio™ 3D Digital PCR Sample Loading Blade.
2. Plug in and power on the QuantStudio™ 3D Digital PCR Chip Loader, then wait until the Chip Loader status light illuminates solid green (≥20 min depending on room temperature).
3. Remove the Chip Sealant syringe, plunger, and tip from the protective packaging. Remove the protective caps from both ends of the syringe, twist and push the tip to lock it into place, then insert the plunger into the opposite end of the syringe. Syringe can be stored and reused (*see* **Note 25**).
4. Remove the QuantStudio™ 12 K Flex OpenArray® Immersion Fluid syringe, plunger, and tip from the packaging. Unscrew the cap from the syringe, then attach the OpenArray® Immersion Fluid Tip by pushing it into place and confirm tip locked in place before proceeding. Carefully depress the plunger until Immersion Fluid flows from the tip of the assembled syringe. Use within 1 h of opening (*see* **Note 26**).
5. Open the QuantStudio™ 3D Digital PCR 20 K Chip package, then gently grasp the chip by its sides and load it face-up into the chip nest. Lock the chip into place by pressing down the chip nest lever prior to placing the chip into the chip nest.
6. Briefly vortex and centrifuge the prepared digital PCR reaction, then carefully transfer 16.7 μL of the solution into the sample-loading port of the QuantStudio™ 3D Digital PCR Sample Loading Blade. Press the sample loading blade lever, then install the Sample Loading Blade to the Chip Loader.

7. Remove the red protective film from a QuantStudio™ 3D Digital PCR Chip Case Lid, press the Lid Nest button, carefully place the Chip Case Lid into the Lid Nest with the barcode on the underside, and the Immersion Fluid opening on the bottom right. Then release the button to clamp the chip in place.
8. Press the black loading button on the Chip Loader to load the Digital PCR 20K Chip. The status light flashes green during the loading sequence, and displays solid green when finished. After the Chip Loader loads the Digital PCR 20 K Chip, immediately add 10–15 drops of Immersion Fluid directly onto the chip so that the fluid covers the entire surface (*see* **Note 27**).
9. Rotate the Chip Loader arm so that the Chip Case Lid solidly contacts the Digital PCR 20K Chip, firmly press down for 15 s to ensure a tight seal, then press the Lid Nest button and return the Chip Loader arm to its original position (*see* **Note 28**).
10. Fill the Chip Case with Immersion Fluid by holding the Chip Case by its edges and at a 45° angle so that air can escape from the loading port as you fill it. Fill until the Chip Case contains an air bubble no larger than the fill port (<2–3 mm in diameter). Check for air bubbles that might be hidden behind the serial number label. Remove any excess Immersion Fluid from the Chip Case to ensure optimal imaging on the QuantStudio™ 3D Instrument.
11. Seal the Chip Case using Chip Sealant by inserting the syringe tip into the fill port of the sealed Chip Case, then carefully fill the port with Chip Sealant, ensuring that the fluid touches the walls of the hole. When full, finish the seal by lifting the syringe tip and add a small amount of sealant over the top of the port.
12. Insert the Digital PCR 20K Chip assembly into the UV-Curing Station on the Chip Loader. The ultraviolet light will illuminate for approximately 15 s, then remove the chip and place it on a clean, dry, lint-free surface (*see* **Note 29**).
13. Visually inspect the sealed Digital PCR 20K Chip for leaks, bubbles. Store the prepared Digital PCR 20K Chip in a clean, dry, dark location and begin thermal cycling within 2 h.

3.3.3 Thermal Cycle the Digital PCR 20K Chips

1. Using the GeneAmp® PCR System 9700, open the heated cover, then wipe the surface of both sample blocks using a lint-free wipe to ensure that they are clean and dry.
2. Confirm the Tilt Base is installed to the thermal cycler and that Chip Adapters are installed to both sample blocks.
3. Load the QuantStudio™ 3D Digital PCR 20K Chips into the thermal cycler with the fill ports on the chips positioned toward

Table 5
PCR cycling conditions for the digital PCR 20 K chips

PCR protocol						
Stage 1	**Stage 2**		**Stage 3**		**Run speed**	**Rxn. vol.**
96.0 °C 10:00 1×	60.0 °C 2:00 39×	98.0 °C 0:30	60.0 °C 2:00 1×	10.0 °C 99:59	Std.	20 μL

the front of the thermal cycler. Lay the QuantStudio™ Thermal Pads over the Digital PCR 20K Chips. Slide the heated cover forward and pull the heated cover lever down to engage the cover with the QuantStudio™ 3D Digital PCR 20K Chips (*see* **Note 30**).

4. Use the thermal cycler to select and start the preprogrammed run (Table 5) for the Digital PCR 20K Chips.
5. Remove the Digital PCR 20K Chips from the Chip Adapters and allow them to equilibrate to room temperature. The Digital PCR 20K Chip is now ready to be imaged by the QuantStudio™ 3D Instrument within 1 h of removal from the thermal cycler (*see* **Note 31**).

3.3.4 Analyzing the Digital PCR 20K Chips

1. In the Destination screen from the Main Menu of the QuantStudio™ 3D Instrument touch the desired destination for the imaging data. If using use a USB drive, insert the drive into the USB port on the front of the instrument.
2. Open the chip tray and load the Digital PCR 20K Chip into the bay. Confirm that the Digital PCR 20K Chip is correctly aligned with the barcode on top and facing the user.
3. In the instrument touchscreen, touch Start Run, then wait for the QuantStudio™ 3D Instrument to image the chip (*see* **Note 32**).
4. When the touchscreen displays the Analyzing Chip screen remove the Digital PCR 20K Chip. Store or discard it as desired (*see* **Note 33**).
5. After reviewing the results of the run, touch Done to close the results screen (*see* **Note 34**).

4 Notes

1. We have regularly extracted exosomal RNA from samples stored in PBS or samples collected in Optiprep© Gradient fractions using TRIzol® LS.

2. It is essential that TRIzol® LS is used if exosomes are resuspended in a buffer such as PBS or Optiprep© Gradient. TRIzol® LS is a concentrated version of TRIzol® and ideal for samples resuspended in buffer. TRIzol® should only be used for cell pellets or tissue samples which are not suspended in buffer.
3. Exosomes resuspended in 250 μL of buffer must maintain the correct sample to TRIzol® LS and chloroform ratio to yield maximal RNA and quality.
4. After centrifugation, the sample separates into three phases: an upper, colorless, aqueous phase containing RNA; a white interphase; and a lower, red, organic phase. Following centrifugation, if the interphase layer is not compact, repeat the centrifugation step.
5. The aqueous layer from the RNA extraction must be accurately measured in order to determine the correct volume of 100% ethanol to precipitate RNA. The 1.5 volumes of ethanol ensure precipitation of small RNAs and miRNAs.
6. Optional QIAcube for automation miRNeasy mini kit can be continued from this step following manufacturer's instructions.
7. The long centrifugation dries the spin column membrane, ensuring that no ethanol is carried over during RNA elution. Residual ethanol may interfere with downstream reactions. Following centrifugation, remove the miRNeasy Mini spin column from the collection tube carefully so the column does not contact the flow-through. Otherwise, carryover of ethanol will occur.
8. Perform this step to eliminate any possible carryover of Buffer RPE or if residual flow-through remains on the outside of the miRNeasy Mini spin column after **step 15**.
9. If the expected RNA yield is >30 μg, repeat **step 18** with a second volume of 30–50 μL RNase-free water. Elute into the same collection tube. To obtain a higher total RNA concentration, this second elution step may be performed by using the first eluate (from **step 19**). The yield will be 15–30% less than the yield obtained using a second volume of RNase-free water, but the final concentration will be higher.
10. MicroRNA RT reactions can be multiplexed with up to 96 individual 5× primers and and/or TaqMan MicroRNA Assays for with or without preamplification in each reaction without loss of performance [22]. We have successfully multiplexed up to 20 assays per RT reaction following the protocol with or without preamplification.
11. Assessment of RNA quality and integrity is crucial to successful downstream applications for gene expression analysis as

microRNA expression cannot be accurately profiled from degraded total RNA samples. Running samples on the Bioanalyzer with the RNA 6000 kit provides an accurate assessment of both the RNA quantity and quality by determining the RNA integrity number (RIN) of the given sample. A RIN greater than or equal to 7 is recommended for the profiling of microRNAs [23].

12. Centrifugal force greater than 1000 × *g* may result in rubber caps of 5× RT primer and 20× assays tubes being lodged inside tubes.
13. Do not vortex MultiScribe™ Reverse Transcriptase or RT master mix containing MultiScribe™ Reverse Transcriptase.
14. Include additional reactions in the calculations to provide 20% excess volume for the loss that occurs during reagent transfers.
15. For each microRNA assay an individual reverse transcription (RT) reaction must be prepared per sample. Therefore the final 15 μL reaction contains 7 μL RT master mix: 5 μL RNA: and 3 μL of 5× RT primer specific for each individual miRNA.
16. Keep all TaqMan reagents protected from light, in the freezer, until you are ready to use them. Excessive exposure to light may affect the fluorescent probes.
17. The final diluted cDNA template volume is sufficient for eight replicates.
18. We recommend quadruplicate replicates for each assay; however, triplicate replicates may be performed if necessary. Calculating the total volume required for each component by the number of replicates for each sample according to the table. For every four reactions, include volume for a fifth reaction to provide excess volume for the loss that occurs during reagent transfers.
19. Optional, reaction plate can be setup on QIAgility or equivalent liquid handling platform according to manufacturer's instructions.
20. If using Taqman Gene expression master mix run the plate in standard mode. Check thermal-cycling parameters of your instrument with master mix compatibility as parameters may vary.
21. A universally excepted endogenous control for microRNA has yet to be established. Based on our experience we find the RNU6B, snoRNA202 and snoRNA135 to show the least variability between mouse hypothalamic GT1-7 samples [9].
22. It is recommended that a set of several endogenous small nuclear RNA (snRNA) and/or small nucleolar RNA (snoRNA) endogenous controls control genes be selected and screened based on the species, tissues, or cell lines used in your study. Alternatively, or in addition to, use specific miRNAs

that demonstrate the least variability across experimental conditions. Applied Biosystems recommends the following candidate control genes for Human: RNU48, RNU44, U47, RNU6B and Mouse: snoRNA202, snoRNA234 samples based upon data generated by across a wide variety of tissues and cell lines that show the least variability [24]. Alternatively, the use of a spike-in-control such as *C. elegans* miR-39 may be used to normalize [14].

23. The recommended cDNA volume is based upon a human gDNA sample at 10 ng/μL concentration with the target sequence present at two copies per diploid genome. This recommended volume will vary depending upon species, sample type, and sample concentration. For miRNA samples we have found that using either neat undiluted cDNA, 1:5 diluted or 1:10 diluted gives optimal results for target sequence between 200 and 2000 copies/μL. As a general guide, if the Ct value of the target miRNA is ~15–20 then 1:10 dilution is recommended; Ct value of 20–25 then 1:5 is recommended and Ct value >25 then undiluted neat cDNA is recommended.
24. For optimal results, load the Digital PCR 20K Chips as soon as possible after setting up the reactions. If necessary, place the reactions on ice until you are ready to use them. If placed on ice, warm the samples to room temperature prior to loading.
25. Do not discard the Chip sealant packaging. When not in use store the Chip sealant in its protective packaging.
26. Use all of the immersion fluid within 1 h of uncapping syringe and then dispose of. Once a syringe is opened, you cannot recap for later use.
27. If any fluid is present on the edges of the case that will contact the lid, remove it with a lint-free wipe that has been sprayed with isopropanol.
28. Press down on the Chip Loader until you reach a hard stop. The arms requires >20 lbs of force to correctly apply and seal the Chip Case Lid.
29. Do not allow the Digital PCR 20K Chip to contact the roof of the UV-Curing Station.
30. Load the right sample block first, placing at least 1 chip on the right sample block. Balance the load between the left and right sample blocks so that the pressure applied by the heated cover and thermal pads is uniform across all samples.
31. Visually inspect each Digital PCR 20K chip surface and clean using a lint-free laboratory wipe to remove any condensation or Immersion Fluid from the surface of the chip by wiping in one direction. If necessary, use a lint-free laboratory wipe sprayed with isopropanol to remove any dried residue.

32. Do not open the tray or remove the USB drive while the instrument displays the countdown screen. Doing so will invalidate the image data and require you to repeat the run.
33. You can read another Digital PCR 20K Chip without waiting for the analysis to complete. To begin the next run, load the chip and touch Start Run as described in **step 2**. Data automatically saves to the USB. To identify samples, run the samples in order and arrange by the timecode incorporated into the data file name.
34. If the analysis of the imaging data produced a red flag for either probe (FAM™ or VIC® dye), visually inspect the chip for problems and read the chip again.

References

1. Bartel DP (2004) MicroRNAs: genomics, biogenesis, mechanism, and function. Cell 116:281–297
2. Griffiths-Jones S (2004) The microRNA registry. Nucleic Acids Res 32:D109–D111
3. Vidigal JA, Ventura A (2015) The biological functions of miRNAs: lessons from in vivo studies. Trends Cell Biol 25:137–147
4. Cheng L, Sharples RA, Scicluna BJ et al (2014) Exosomes provide a protective and enriched source of miRNA for biomarker profiling compared to intracellular and cell-free blood. J Extracell Vesicles 3:PMCID: PMC3968297
5. Cheng L, Sun X, Scicluna BJ et al (2014) Characterization and deep sequencing analysis of exosomal and non-exosomal miRNA in human urine. Kidney Int 86:433–444
6. Valadi H, Ekstrom K, Bossios A et al (2007) Exosome-mediated transfer of mRNAs and microRNAs is a novel mechanism of genetic exchange between cells. Nat Cell Biol 9:654–659
7. Andreasen N, Blennow K (2005) CSF biomarkers for mild cognitive impairment and early Alzheimer's disease. Clin Neurol Neurosurg 107:165–173
8. Murata K, Yoshitomi H, Tanida S et al (2010) Plasma and synovial fluid microRNAs as potential biomarkers of rheumatoid arthritis and osteoarthritis. Arthritis Res Ther 12:R86
9. Bellingham SA, Coleman BM, Hill AF (2012) Small RNA deep sequencing reveals a distinct miRNA signature released in exosomes from prion-infected neuronal cells. Nucleic Acids Res 40:10937–10949
10. Kalra H, Adda CG, Liem M et al (2013) Comparative proteomics evaluation of plasma exosome isolation techniques and assessment of the stability of exosomes in normal human blood plasma. Proteomics 13:3354–3364
11. Croce CM (2009) Causes and consequences of microRNA dysregulation in cancer. Nat Rev Genet 10:704–714
12. Small EM, Olson EN (2011) Pervasive roles of microRNAs in cardiovascular biology. Nature 469:336–342
13. Eacker SM, Dawson TM, Dawson VL (2009) Understanding microRNAs in neurodegeneration. Nat Rev Neurosci 10:837–841
14. Cheng L, Doecke JD, Sharples RA et al (2015) Prognostic serum miRNA biomarkers associated with Alzheimer's disease shows concordance with neuropsychological and neuroimaging assessment. Mol Psychiatry 20(10):1188–1196
15. Rabinowits G, Gercel-Taylor C, Day JM et al (2009) Exosomal microRNA: a diagnostic marker for lung cancer. Clin Lung Cancer 10:42–46
16. Taylor DD, Gercel-Taylor C (2008) MicroRNA signatures of tumor-derived exosomes as diagnostic biomarkers of ovarian cancer. Gynecol Oncol 110:13–21
17. Nilsson J, Skog J, Nordstrand A et al (2009) Prostate cancer-derived urine exosomes: a novel approach to biomarkers for prostate cancer. Br J Cancer 100:1603–1607
18. Van Balkom BW, Pisitkun T, Verhaar MC et al (2011) Exosomes and the kidney: prospects for diagnosis and therapy of renal diseases. Kidney Int 80:1138–1145
19. Coleman BM, Hanssen E, Lawson VA et al (2012) Prion-infected cells regulate the release of exosomes with distinct ultrastructural features. FASEB J 26:4160–4173

20. Cheng L, Quek CY, Sun X et al (2013) The detection of microRNA associated with Alzheimer's disease in biological fluids using next-generation sequencing technologies. Front Genet 4:150
21. Pfaffl MW (2001) A new mathematical model for relative quantification in real-time RT-PCR. Nucleic Acids Res 29, e45
22. Appliedbiosystems (2013) Procedure for multiplexing the RT step with or without preamplification while using TaqMan MicroRNA assays. In: Applied biosystems user bulliten 4465407. http://www.appliedbiosystems.com
23. Ibberson D, Benes V, Muckenthaler MU et al (2009) RNA degradation compromises the reliability of microRNA expression profiling. BMC Biotechnol 9:102
24. Wong L, Lee K, Russell I et al (2007) Endogenous controls for real-time quantitation of miRNA using TaqMan® MicroRNA assays. In: Applied biosystems application note. http://www.appliedbiosystems.com

Chapter 6

Small RNA Library Construction for Exosomal RNA from Biological Samples for the Ion Torrent PGM™ and Ion S5™ System

Lesley Cheng and Andrew F. Hill

Abstract

Next-generation deep sequencing (NGS) technology represents a powerful and innovative approach to profile small RNA. Currently, there are a number of large-scale and benchtop sequencing platforms available on the market. Although each platform is relatively straightforward to operate, constructing cDNA libraries can be the most difficult part of the NGS workflow. Constructing quality libraries is essential to obtaining a successful sequencing run of high-quality reads and coverage. The quality and yield of RNA affect hybridization and ligation of sequencing adapters. In the field of biomarker discovery, there has been an interest in profiling exosomal RNA from biological fluids. However, very little RNA yield is obtained when extracting RNA from exosomes, thus making library construction difficult. Here, this protocol describes an optimized protocol for constructing small RNA libraries from low yields of RNA, in particular, extracted from exosomes isolated from biological fluids.

Key words Small RNA deep sequencing, Ion Torrent, Exosomes, Biological fluids, miRNA, Small RNA

1 Introduction

Next-generation deep sequencing (NGS) has allowed the ability to profile and discover noncoding small RNA species. One of the most studied species of small RNA is microRNA (miRNA) of approximately 22–24 nucleotides which acts as a post-translational regulator of gene expression [1]. Consequently, miRNA can regulate essential biological pathways such as cellular development, proliferation, apoptosis, and cellular signaling. Currently there are more than 2500 known miRNA species with each having multiple messenger RNA gene targets [2]. As they are highly abundant in cells and tissues, extracting small RNA including miRNA is relatively uncomplicated, and small RNA libraries can be successfully constructed. However, miRNA is also found as smaller quantities

Andrew F. Hill (ed.), *Exosomes and Microvesicles: Methods and Protocols*, Methods in Molecular Biology, vol. 1545, DOI 10.1007/978-1-4939-6728-5_6, © Springer Science+Business Media LLC 2017

in biological fluids such as blood, urine, cerebral spinal fluid, and synovial fluid [3–6].

RNA extraction from these biological fluids can be difficult as there is a high level of protein contaminates such as albumin and immunoglobulins. In addition, there is little RNA circulating in these fluids due to the presence of RNases. However, miRNA can be secreted extracellularly in cell-derived extracellular vesicles such as exosomes where they are protected from RNases and remain relatively stable. miRNAs have been implicated in many diseases and have been shown to be taken up by distant cells as cargo in exosomes as a method of cell-to-cell communication to potentially influence disease pathogenesis and progression [7]. Together, the pathogenic nature of miRNAs and ability to be secreted extracellularly into biological fluids, this presents miRNA as a promising biomarker for diagnosis and therapeutic monitoring for diseases such as cancer, neurodegenerative disorders, heart disease, and infection.

Firstly, exosomes can be isolated from biological fluids by sequential ultracentrifugation which involves the removal of large debris, proteins, and microvesicles through low-speed centrifugation [8]. This is followed by pelleting the smaller membrane vesicles, exosomes, by high-speed ultracentrifugation as described elsewhere. Isolation of exosomes from various biological fluids can be challenging in addition to extracting the RNA contained in the exosomes (termed exosomal RNA). The minimal yields obtained of exosomal RNA consequently makes constructing small RNA libraries difficult.

The team at Ion Torrent has developed a kit for small RNA libraries (Ion Total RNA-Seq Kit v2) mostly catered for RNA-rich samples extracted from cells and tissues. Here, the protocol adapted from the Ion Total RNA-Seq Kit v2 has been modified for samples with less than 10 ng of RNA. In particular, the protocol outlined here is for exosomes isolated from biological fluids which contain an enriched population of miRNA and small RNA species [3, 9]. Upon constructing the small RNA libraries, libraries can be templated using the One Touch 2 or Ion Chef system and sequenced on the Ion Torrent Personal Genome Machine (PGM™) or Ion S5 benchtop sequencer with the potential to obtain 5–6 million reads per 318™ chip or 80 million reads per 540 chip using 200 bp sequencing.

2 Materials

2.1 RNA Isolation and Quantification

1. Phosphate-buffered saline (PBS).
2. Screw cap tubes (1.8 ml).
3. miRNeasy mini kit (Qiagen).

4. TRIzol® LS reagent (Life Technologies).
5. Chloroform.
6. Nonstick Lo-bind RNase-free microfuge tubes (1.5 ml).
7. Ethanol, 100 %, ACS reagent grade or equivalent.
8. Nuclease-free water.
9. Small RNA Bioanalyzer® assay kit (Agilent).
10. RNA Nano or Pico Bioanalyzer® assay kit (Agilent).
11. Bioanalyzer® instrument (Agilent).

2.2 Enrichment of Small RNA

1. Ion Total RNA-Seq Kit v2 components: 96-well deep well processing plate, nucleic acid binding beads, binding solution, and washing solution with ethanol added as indicated.
2. Ethanol, 100 %, ACS reagent grade or equivalent.
3. 96-well magnetic stand.
4. Heat block to 80 °C.
5. Nonstick RNase-free microfuge tubes (1.5 ml).
6. Small RNA Bioanalyzer® assay kit.
7. RNA Nano or Pico Bioanalyzer® assay kit.

2.3 Hybridize and Ligate RNA

1. PCR RNase-free 0.2-ml tubes.
2. Ion Total RNA-Seq Kit v2 components: Ion Adapter Mix v2, hybridization solution, 2× ligation buffer, and ligation enzyme buffer.
3. Thermal cycler with heated lid, for 0.2-ml tubes.

2.4 Synthesis of cDNA for PCR Amplification

1. Nuclease-free water.
2. Ion Total RNA-Seq Kit v2 components: 10× RT buffer, 2.5 nM dNTP mix, Ion RT Primer v2, 10× SuperScript® III enzyme mix.
3. PCR RNase-free 0.2-ml tubes.
4. Thermal cycler with heated lid, for 0.2-ml tubes.

2.5 Purify and Size Select cDNA Products

1. Ion Total RNA-Seq Kit v2 components: 96-well deep well processing plate, nucleic acid binding beads, binding solution, and washing solution with ethanol added as indicated.
2. Ethanol, 100 %, ACS reagent grade or equivalent.
3. 96-well magnetic stand.
4. Heat block to 37 °C.
5. Nonstick Lo-bind RNase-free microfuge tubes (1.8 ml).

2.6 Amplify the cDNA by PCR

1. Ion Total RNA-Seq Kit v2 components: Platinum® PCR SuperMix High Fidelity, Ion 5′ PCR primer v2, and Ion 3′ PCR primer v2.

2. Ion Xpress RNA 3′ barcode primer.
3. Ion Xpress RNA-Seq Barcode BC primer (BC01–BC16).
4. Thermal cycler.

2.7 Purify and Size Select PCR Products

1. Ion Total RNA-Seq Kit v2 components: 96-well deep well processing plate, nucleic acid binding beads, binding solution, and washing solution with ethanol added as indicated.
2. Ethanol, 100%, ACS reagent grade or equivalent.
3. 96-well magnetic stand.
4. Heat block to 37 °C.
5. Nonstick Lo-bind RNase-free microfuge tubes (1.8 ml).

2.8 Assess the Yield and Size Distribution of the Amplified Library cDNA

1. DNA 1000 Bioanalyzer® assay kit (Agilent).
2. Bioanalyzer® instrument (Agilent).
3. Low Tris-EDTA buffer (supplied in Bioanalyser kit).
4. Nonstick Lo-bind RNase-free microfuge tubes (1.8 ml).

3 Methods

3.1 RNA Extraction of Exosomal Samples

It is essential that handling and collection procedures for biological samples are carried out according to strict protocols to ensure RNA does not degrade. In addition, exosome isolation should also be optimized with the aim to collect maximal yields of exosomes from the fluid [10, 11]. *See* **Note 1**.

1. Cool centrifuge to 4 °C.
2. Ensure exosome sample has been resuspended in 250 μl of PBS within a screw cap microfuge tube (1.8 ml).
3. Add 750 μl of TRIzol® LS, vortex for 5–10 s to ensure proper lysis, and incubate at room temperature for 5 min. *See* **Note 2** for more information.
4. Add 250 μl of chloroform, secure the cap, and shake for 15 s. Incubate at room temperature for 3 min. *See* **Note 3** for more information.
5. Centrifuge at 12,000 × *g* for 15 min at 4 °C.
6. Transfer the upper aqueous layer into a Lo-bind DNA microfuge tube. Do not transfer or disrupt the interphase as this will introduce protein and DNA contamination.
7. Measure the volume of aqueous phase and add exactly 1.5 volumes of 100% fresh ethanol to precipitate small and large RNA. Mix thoroughly by pipetting.
8. Transfer 700 μl of sample into an RNeasy mini column in a 2-ml collection tube provided in the miRNeasy mini kit.

9. Close the lid of the column and centrifuge at 8000 × *g* for 15 s at room temperature. Discard the flow through.
10. Repeat using the remainder of the sample.
11. Add 700 μl of Buffer RWT to the RNeasy mini column. Close the lid, and centrifuge at 8000 × *g* for 15 s at room temperature. Discard flow through.
12. Add 500 μl of Buffer RPE to the RNeasy mini column. Close the lid, and centrifuge at 8000 × *g* for 15 s at room temperature. Discard flow through.
13. Add 500 μl of Buffer RPE to the RNeasy mini column. Close the lid, and centrifuge at 8000 × *g* for 2 min at room temperature. Discard flow through.
14. Transfer the RNeasy mini column into a new 2-ml collection tube. Centrifuge the column with its lid open at 13,000 × *g* for 1 min at room temperature to ensure the membrane is dry and devoid of ethanol.
15. Transfer the RNeasy mini column to a new 1.5-ml collection tube for elution of RNA.
16. Add 100 μl of RNase-free water directly onto the RNeasy mini column membrane.
17. Incubate for 1 min at room temperature.
18. Close the lid of the column and centrifuge at 8000 × *g* for 1 min at room temperature to elute the RNA.
19. Concentrate the sample from 100 to 6 μl using a centrifugal vacuum concentrator (e.g., SpeedyVac®). *See* **Note 4**.
20. Thoroughly resuspend the concentrated RNA in the new volume of 6 μl and centrifuge the sample to the bottom of the tube.
21. Use 1 μl of the exosomal RNA sample to quantitate and assess the small RNA profile using a small RNA Bioanalyzer® assay. Follow the manufacturer's instructions for performing the assay.
22. Use 1 μl of the exosomal RNA sample to quantitate and assess the large RNA profile using a RNA Nano or Pico Bioanalyzer® assay. Follow the manufacturer's instructions for performing the assay.

3.2 Assessing Exosomal RNA Profiles and Yield

The ultimate success of library construction can be largely reliant on the quality and profile of the exosomal RNA samples. The majority of exosomes isolated from biological fluids are enriched with small RNA of less than 200 nt with minimal or no large RNA species (>200 nt). The absence of large ribosomal RNA makes assessing the RNA integrity number (RIN) not possible or challenging. However, a total RNA analysis (0–4000 nt) should be performed if the sample is suspected of containing RNA species larger than 200 nt. The aim of assessing the small RNA profile is to determine the percentage of miRNA in your sample. The sample must contain greater than 1 % miRNA to ensure successful library construction.

1. Upon running the exosomal RNA sample on a small RNA chip, determine the mass of total RNA and miRNA (10–40 nt).
   (a) If the sample has small and large RNA species, calculate the percentage of miRNA using this formula:
   % miRNA = (mass of miRNA (ng) ÷ mass of total RNA) × 100.
   (b) If the sample has only small RNA and no large RNA species, calculate the percentage of miRNA using this formula:
   % miRNA = (mass of miRNA ÷ mass of total small RNA) × 100.
2. Determine whether the sample requires small RNA enrichment:

≥1 % miRNA	Small RNA enrichment is not required, and the RNA sample can be used in the ligation reaction. Continue to "hybridize and ligate RNA."
	However, if the sample contains significant large ribosomal RNA species, it is still recommended to perform small RNA enrichment for optimal ligation of sequencing adapters. Therefore, continue to "enrichment of small RNA" (Fig. 1).
≤1 % miRNA	Small RNA enrichment is required. Continue to "enrichment of small RNA."

3.3 Enrichment of Small RNA

The enrichment of small RNA assists in the ligation of sequencing adapters to only the small RNA of interest rather than larger RNA fragments that may be present in the sample, although, the observation that exosomes isolated from biological fluids only contain small RNA consequently has been biologically enriched for small RNA. Thus, enrichment by nucleic acid beads would not be required for samples biologically enriched with small RNA.

1. Pre-warm nuclease-free water to 80 °C.
2. Gently vortex the nucleic acid binding beads to resuspend the beads.
3. Add 7 μl of beads to the wells of the processing plate.
4. Accurately add 120 μl binding solution concentrate to each well and mix by pipetting up and down ten times. Avoid bubbles being generated to ensure correct measurement of 120 μl.
5. Resuspend RNA sample (maximum of 20 μg total RNA) in 75 μl nuclease-free water.

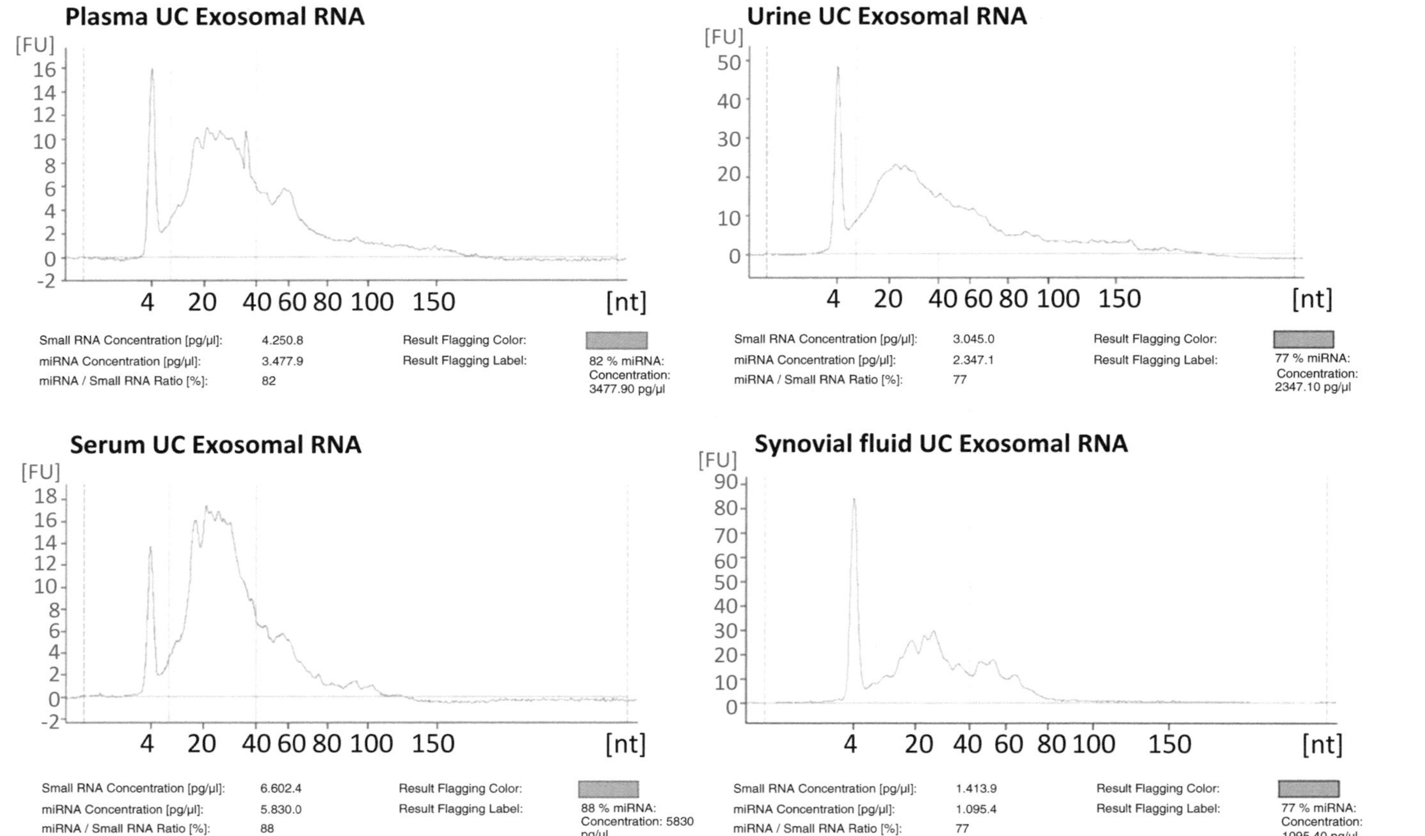

Fig. 1 Small RNA profiles of exosomes secreted from various biological fluids. Exosomes were isolated from 1 ml of plasma and serum, 20 ml of urine, and 5 ml of synovial fluid using sequential ultracentrifugation (UC). Samples were concentrated on a SpeedVac® instrument to 6 µl. RNA profiles were analyzed using a small RNA assay kit and ran on an Agilent® Bioanalyzer® Instrument. A ladder was run during each Bioanalyzer® run which was used to calibrate the fluorescence units (FUs), nucleotide (nt) size, and RNA yield which may differ from chip to chip. These samples did not contain detectable large RNA species (≥200 nt) as examined on a RNA Pico Bioanalyzer® chip. Therefore, the small RNA and miRNA (10–40 nt) concentration was used to calculate percentage of miRNA and miRNA yield (ng)

6. Transfer 75 μl of each RNA sample to a well containing beads in the processing plate.
7. Set a pipette at 105 μl and attach a new 200 μl tip to the pipette. Pre-wet the tip with 100% ethanol by pipetting the ethanol up and down three times.
8. Without changing the pre-wet tip, add 105 μl of 100% ethanol to each well. Upon dispensing the ethanol, do not pipette to the last pipette stop or dispense the last drop. Rather, remove the last drop by touching the tip to the well wall. This allows for accurate pipetting of 100% ethanol, thus providing the best results for size selection. Continue with the same pre-wet tip for each well. If you have contaminated the pre-wet tip with RNA sample, prepare another pre-wet tip.
9. With a new pipette tip for each well, resuspend the suspension in each well by pipetting the wells up and down ten times. The suspension should be homogenous in color after mixing.
10. Incubate the suspension for 5 min at room temperature to allow large RNA fragments to bind to the beads.
11. Place the processing plate on the magnetic stand for 5 min to separate the beads from solution. Ensure the supernatant clears before proceeding.
12. While the processing plate remains on the magnetic stand, transfer the supernatant to a new well on the plate. The supernatant contains the small RNA species.
13. Remove the processing plate from the magnetic stand.
14. Add 30 μl of nuclease-free water to the supernatant transfered in step 12.
15. Set a pipette at 570 μl and attach a new 1000 μl tip to the pipette. Pre-wet the tip with 100% ethanol by pipetting the ethanol up and down three times.
16. Without changing the pre-wet tip, add 570 μl of 100% ethanol to each well.
17. Gently vortex the nucleic acid binding beads tube to resuspend the beads.
18. Add 7 μl beads to the wells of the processing plate.
19. Set a P200 pipette at 150 μl and attach a new 200 μl tip to the pipette. Resuspend the suspension in each well by pipetting the wells up and down ten times. The suspension should be homogenous in color after mixing.
20. Incubate the suspension for 5 min at room temperature to allow small RNA fragments to bind to the beads.
21. Place the processing plate on the magnetic stand for 5 min to separate the beads from solution. Ensure the supernatant clears before proceeding.

22. Leave the processing plate on the magnetic stand, then carefully aspirate and discard the supernatant from the plate.
23. While the processing plate remains on the magnetic stand, add 150 μl wash solution concentrate (with ethanol) to each sample.
24. Incubate on the magnetic stand for 30 s. Do not resuspend beads.
25. Leave the processing plate on the stand, and then carefully aspirate and discard the supernatant from the plate.
26. Remove all of the wash solution concentrate from the well without disturbing the magnetic beads in the well. Use a 10 μl pipette to remove residual ethanol.
27. Air-dry the beads at room temperature for 1–2 min. Do not over dry the beads. Over-dried beads appear cracked.
28. Remove the processing plate from the magnetic stand.
29. Elute the small RNA with 30 μl of pre-warmed (80 °C) nuclease-free water to each sample.
30. Mix thoroughly by pipetting up and down ten times.
31. Incubate the sample for 1 min.
32. Place the processing plate on the magnetic stand for 1 min to separate the beads from solution.
33. Ensure the solution has cleared before proceeding.
34. For each sample, collect 30 μl of eluent.
35. Run 1 μl of the enriched small RNA sample on the Agilent® Bioanalyzer® using a small RNA assay kit. Follow the manufacturer's instructions.

3.4 Hybridize and Ligate RNA

One of the essential steps to constructing libraries for deep sequencing is to ligate sequencing adapters to RNA fragments in the sample. These adapters may be platform specific and include a set of single-stranded oligonucleotides with a defined sequence used to identify the start and end of a RNA fragment for sequencing. The Ion Adapter Mix v2 contains a 5′ primer and 3′ primer for ligation.

Table 1
Determining the amount of RNA required

Sample type	Amount of miRNA (10–40 nt) in 3 μl	Total RNA of sample, μg
Sample contains small and large RNA species	5–100 ng	≤1
Sample enriched with miRNA and small RNA only	1–100 ng Ideally, use ≥10 ng of miRNA	≤1

Refer to Fig. 1 for examples of samples containing enriched small RNA

1. Using the results from the Agilent® 2100 Bioanalyzer® instrument and the small RNA assay kit, determine the amount of total RNA (in particular miRNA) to use according to the RNA profile of the sample (Table 1). The amount of total RNA needs to be in a volume of 3 μl. If necessary, concentrate the small RNA with a SpeedVac® centrifugal concentrator. *See* **Notes 5** and **6**.
2. Add 3 μl of the small RNA sample as determined above into a 0.2-ml PCR tube.
3. Add 3 μl of hybridization solution to each sample.
4. Add 2 μl of Ion Adapter Mix v2 to each sample. The total volume per reaction should be 8 μl.
5. Pipet the mixture up and down five times to mix, then centrifuge briefly to collect the liquid at the bottom of the tube.
6. Run the hybridization reaction in a thermal cycler according to Table 2 with the lid temperature set at 100 °C.
7. While the hybridization reaction occurs, thaw out the 2× ligation buffer, and briefly centrifuge down any white precipitate found inside the lid of the tube. Warm the 2× ligation buffer at 37 °C, and vortex until any precipitate until dissolved.
8. Once the hybridization reaction has completed, place the reactions on ice.
9. Add 10 μl of 2× ligation buffer to each sample.
10. Add 2 μl of ligation enzyme mix to each sample. The total volume of the reaction should now be 20 μl.
11. Pipet the mixture up and down five times to mix, then centrifuge briefly to collect the liquid at the bottom of the tube.
12. Incubate the 20 μl ligation reaction in a thermal cycler at 16 °C for 16 h with the thermal cycler lid set to 16 °C.

3.5 Synthesis of cDNAs for PCR Amplification

Once the sequencing adapters have been ligated on both 5′ and 3′ ends of RNA fragments present in the sample, they are then used to perform a reverse transcription to generate single-stranded cDNA of the RNA fragments.

1. Place the overnight ligation reactions on ice, and prepare the samples for reverse transcription accordingly to Table 3 or 4:

Table 2
Thermal cycle conditions for hybridization

Temperature, °C	Time, min
65	10
16	5

Table 3
cDNA reaction for miRNA input ≤10 ng

Component	Volume for one reaction, μl
Nuclease-free water	5
10× RT buffer	4
2.5 nM dNTP mix	2
Ion RT Primer v2	5

Table 4
cDNA reaction for miRNA input ≥ 10 ng

Component	Volume for one reaction, μl
Nuclease-free water	2
10× RT buffer	4
2.5 nM dNTP mix	2
Ion RT Primer v2	8

2. Vortex the RT reaction with ligated RNA sample to mix, and centrifuge to collect the liquid in the bottom of the tube. The reaction should now have a total volume of 40 μl.
3. Incubate the RT reaction with ligated RNA sample at 70 °C for 10 min (with a heated lid set at 100 °C), then snap cool on ice.
4. Add 4 μl of 10× SuperScript® III Enzyme Mix to each ligated RNA sample.
5. Gently vortex to mix thoroughly, then centrifuge briefly.
6. Incubate in a thermal cycler at 42 °C for 30 min (with a heated lid set at 100 °C).
7. The cDNA can be stored at −80 °C or used immediately. *See* **Note 6** for more information.

3.6 Purify and Size Select cDNA Products

The cDNA is then purified in order to remove excess primers, dNTPs, and enzymes. In addition, nucleic acid beads are used to size select cDNA products of approximately 77–200 nt. Adapter dimers with no RNA insert are 77 nt, while anything larger suggests a insert was ligated in between the adapters. This protocol should eliminate cDNA products larger than 200 bp for 200 bp deep sequencing.

1. Pre-warm nuclease-free water to 37 °C.
2. Gently vortex the nucleic acid binding beads to resuspend the beads.

3. Add 7 μl of beads to the wells of the processing.
4. Accurately add 140 μl binding solution concentrate to each well and mix by pipetting up and down ten times. Upon aliquoting, avoid bubbles being generated to ensure correct measurement of 140 μl.
5. Transfer 40 μl of the RT reaction to a well containing beads in the processing plate.
6. Set a P200 pipette at 120 μl and attach a new 200 μl tip to the pipette. Pre-wet the tip with 100% ethanol by pipetting the ethanol up and down three times.
7. Without changing the pre-wet tip, add 120 μl of 100% ethanol to each well. *See* **Note 7**.
8. With a new pipette tip for each well, resuspend the suspension in each well by pipetting the wells up and down ten times. The suspension should be homogenous in color after mixing.
9. Incubate the suspension for 5 min at room temperature to allow large RNA fragments to bind to the beads.
10. Place the processing plate on the magnetic stand for 5 min to separate the beads from solution. Ensure the supernatant clears before proceeding.
11. While the processing plate remains on the magnetic stand, transfer the supernatant to a new well on the plate. The supernatant contains the small RNA species.
12. Remove the processing plate from the magnetic stand.
13. Add 72 μl of nuclease-free water to the supernatant of the new sample well.
14. Set a P200 pipette at 78 μl and attach a new 200 μl tip to the pipette. Pre-wet the tip with 100% ethanol by pipetting the ethanol up and down three times.
15. Without changing the pre-wet tip, add 78 μl of 100% ethanol to each well. *See* **Note 7**.
16. Gently vortex the nucleic acid binding beads tube to resuspend the beads.
17. Add 7 μl beads to the wells of the processing plate.
18. Set a P200 pipette at 150 μl and attach a new 200 μl tip to the pipette. Resuspend the suspension in each well by pipetting the wells up and down ten times. The suspension should be homogenous in color after mixing.
19. Incubate the suspension for 5 min at room temperature to allow small RNA fragments to bind to the beads.
20. Place the processing plate on the magnetic stand for 5 min to separate the beads from solution. Ensure the supernatant clears before proceeding.
21. Leave the processing plate on the magnetic stand, then carefully aspirate and discard the supernatant from the plate.

22. While the processing plate remains on the magnetic stand, add 150 μl wash solution concentrate (with ethanol) to each sample.
23. Incubate on the magnetic stand for 30 s. Do not resuspend beads.
24. Leave the processing plate on the stand, and then carefully aspirate and discard the supernatant from the plate.
25. Remove all of the wash solution concentrate from the well without disturbing the magnetic beads in the well. Use a 10 μl pipette to remove residual ethanol.
26. Air-dry the beads at room temperature for 1–2 min. Do not over dry the beads. Over-dried beads appear cracked.
27. Remove the processing plate from the magnetic stand.
28. Elute the small RNA with 12 μl of pre-warmed (37 °C) nuclease-free water to each sample.
29. Mix thoroughly by pipetting up and down ten times.
30. Incubate the sample for 1 min.
31. Place the processing plate on the magnetic stand for 1 min to separate the beads from solution.
32. Ensure the solution has cleared before proceeding.
33. For each sample, collect 12 μl of eluent.
34. Transfer size selected cDNA into nonstick Lo-bind RNase-free microfuge tubes. cDNA can be stored at −80 °C or used immediately.

3.7 Amplify the cDNA by PCR

1. To prepare non-barcoded libraries, use the primers provided in the Ion Total RNA-Seq Kit v2. To prepare barcoded libraries, use the PCR primers from Ion Xpress™ RNA-Seq Barcode 01–16 Kit. Barcoding libraries enable sequencing of multiple samples in a single multiplexed sequencing run.
2. For each sample, prepare the PCR reactions accordingly in a 0.2-ml PCR tube. Add the reagents in order as displayed in Table 5 or 6.
3. Gently vortex to mix thoroughly, then centrifuge briefly.
4. Run the cDNA sample in a thermal cycler using the conditions in Table 7.

3.8 Purify and Size Select PCR Products

The PCR products are then purified in order to remove excess primers, dNTPs, and enzymes. In addition, nucleic acid beads are used to size select PCR products of approximately 77–200 nt or 85–200 nt for barcoded libraries. This protocol should eliminate cDNA products larger than 200 bp for 200 bp deep sequencing.

1. Pre-warm nuclease-free water to 37 °C.
2. Gently vortex the nucleic acid binding beads to resuspend the beads.

Table 5
Non-barcoded library PCR reactions

Non-barcoded library	
Component	**Volume for one reaction, µl**
Platinum® PCR SuperMix High Fidelity	45
Ion 5′ PCR Primer v2	1
Ion 3′ PCR Primer v2	1
cDNA sample	6
Total volume	53

Table 6
Barcoded library PCR reactions

Barcoded library	
Component	**Volume for one reaction, µl**
Platinum® PCR SuperMix High Fidelity	45
Ion Xpress™ RNA 3′ Barcode Primer	1
cDNA sample	6
Ion Xpress™ RNA-Seq Barcode BC primer (choose from BC01–BC16)	1
Total volume	53

Table 7
PCR cycling conditions

Stage	Temperature, °C	Time
Hold	94	2 min
Cycle (two cycles)	94 50 68	30 s 30 s 30 s
Cycle (14 cycles)	94 62 68	30 s 30 s 30 s
Hold	68	5 min

3. Add 7 µl of beads to the wells of the processing plate.
4. Accurately add 140 µl binding solution concentrate to each well and mix by pipetting up and down ten times.

Upon aliquoting, avoid bubbles being generated to ensure correct measurement of 140 µl.

5. Transfer 53 µl of the PCR reaction to a well containing beads in the processing plate.
6. Set a P200 pipette at 110 µl and attach a new 200 µl tip to the pipette. Pre-wet the tip with 100% ethanol by pipetting the ethanol up and down three times.
7. Without changing the pre-wet tip, add 110 µl of 100% ethanol to each well. *See* **Note** 7.
8. With a new pipette tip for each well, resuspend the suspension in each well by pipetting the wells up and down ten times. The suspension should be homogenous in color after mixing.
9. Incubate the suspension for 5 min at room temperature to allow large RNA fragments to bind to the beads.
10. Place the processing plate on the magnetic stand for 5 min to separate the beads from solution. Ensure the supernatant clears before proceeding.
11. While the processing plate remains on the magnetic stand, transfer the supernatant to a new well on the plate. The supernatant contains the small RNA species.
12. Remove the processing plate from the magnetic stand.
13. Add 35 µl of nuclease-free water to the supernatant of the new sample well.
14. Set a P200 pipette at 35 µl and attach a new 200 µl tip to the pipette. Pre-wet the tip with 100% ethanol by pipetting the ethanol up and down three times.
15. Without changing the pre-wet tip, add 35 µl of 100% ethanol to each well. *See* **Note** 7.
16. Gently vortex the nucleic acid binding beads tube to resuspend the beads.
17. Add 7 µl beads to the wells of the processing plate.
18. Set a P200 pipette at 150 µl and attach a new 200 µl tip to the pipette. Resuspend the suspension in each well by pipetting the wells up and down ten times. The suspension should be homogenous in color after mixing.
19. Incubate the suspension for 5 min at room temperature to allow small RNA fragments to bind to the beads.
20. Place the processing plate on the magnetic stand for 5 min to separate the beads from solution. Ensure the supernatant clears before proceeding.
21. Leave the processing plate on the magnetic stand, then carefully aspirate and discard the supernatant from the plate.
22. While the processing plate remains on the magnetic stand, add 150 µl wash solution concentrate (with ethanol) to each sample.

23. Incubate on the magnetic stand for 30 s. Do not resuspend beads.
24. Leave the processing plate on the stand, and then carefully aspirate and discard the supernatant from the plate.
25. Remove all of the wash solution concentrate from the well without disturbing the magnetic beads in the well. Use a 10 μl pipette to remove residual ethanol.
26. Air-dry the beads at room temperature for 1–2 min. Do not over dry the beads. Over-dried beads appear cracked.
27. Remove the processing plate from the magnetic stand.
28. Elute the small RNA with 10 μl of pre-warmed (37 °C) nuclease-free water to each sample.
29. Mix thoroughly by pipetting up and down ten times.
30. Incubate the sample for 1 min.
31. Place the processing plate on the magnetic stand for 1 min to separate the beads from solution.
32. Ensure the solution has cleared before proceeding.
33. For each sample, collect 10 μl of eluent and store in a nonstick Lo-bind RNase-free microfuge tubes (1.5 ml) at −80 °C until use.

3.9 Assess the Yield and Size Distribution of the Amplified Library cDNA

1. Run 1 μl of the purified cDNA library on an Agilent® 2100 Bioanalyzer® instrument with the Agilent® DNA 1000 Kit or High Sensitivity DNA kit. Follow the manufacturer's instructions for performing the assay.
2. Using the 2100 expert software, perform a smear analysis to determine size distribution of the amplified cDNA:
   (a) Begin with "gating" the size range for all of the ligation products: 50–300 bp.
   (b) Lastly, gate the size range for the desired ligation products.
   For example, to gate miRNA ligated products:
   For non-barcoded libraries: 86–106 bp.
   For barcoded libraries: 94–114 bp.
   Adjust the size range to include all library peaks except for cDNA fragments with no insert (refer to Table 8 and Fig. 2 for examples).
3. Calculate the ratio of the desired ligation products in total ligation products using the formula for:
   Non-barcoded libraries: [Area (86 – desired range bp)] ÷ [Area (50–300 bp)].
   Barcoded libraries: [Area (94 – desired range bp)] ÷ [Area (50–300 bp)].

Table 8

Assess the yield and size distribution of the amplified library cDNA

Insert length, bp	Possible RNA species	Size of non-barcoded library represented on the Bioanalyzer® instrument, bp	Size of barcoded library represented on the Bioanalyzer® instrument, bp
0	No RNA ligated	~77	~85
10	Unknown	~87	~95
20	miRNA	~97	~105
≥50	tRNA, 5s RNA, and other small RNA species	≥127	≥135

If the ratio of the desired products is:

≥50% Proceed to library dilution required for template preparation on the One Touch 2 (approximately 13 pM) or Ion Chef (approximately 45 pM) instrument (*See* **Note 8** and **9**).

≤50% Proceed to library dilution required for template preparation on the One Touch 2 (approximately 13 pM) or the Ion Chef (approximately 45 pM) instrument; however, expect to see an increase in the number of filtered reads (e.g., no insert, excess adapters, etc). Otherwise, start again with a larger input of RNA. *See* **Note 8** and **9**).

4. Determine the molar concentration of cDNA libraries using the size range gated between 50 and 300 bp. Use this concentration to pool barcoded libraries, and determine the library dilution required for template preparation for barcoded and non-barcoded libraries. *See* **Note 10**.
   (a) Continue to template preparation using the One Touch 2 or Ion Chef manual according to the manufacturer's instructions.
   (b) The optimal input of cDNA library for template preparation is between 10-13 pM on the One Touch 2 system or between 43-45 pM on the Ion Chef instrument.

4 Notes

1. It is essential that exosomal isolation procedures are optimized to obtain maximal yields of exosomes from the biological fluid. Consequently, this allows extraction of maximal RNA yield and contributes in the success of constructing small RNA libraries for deep sequencing.

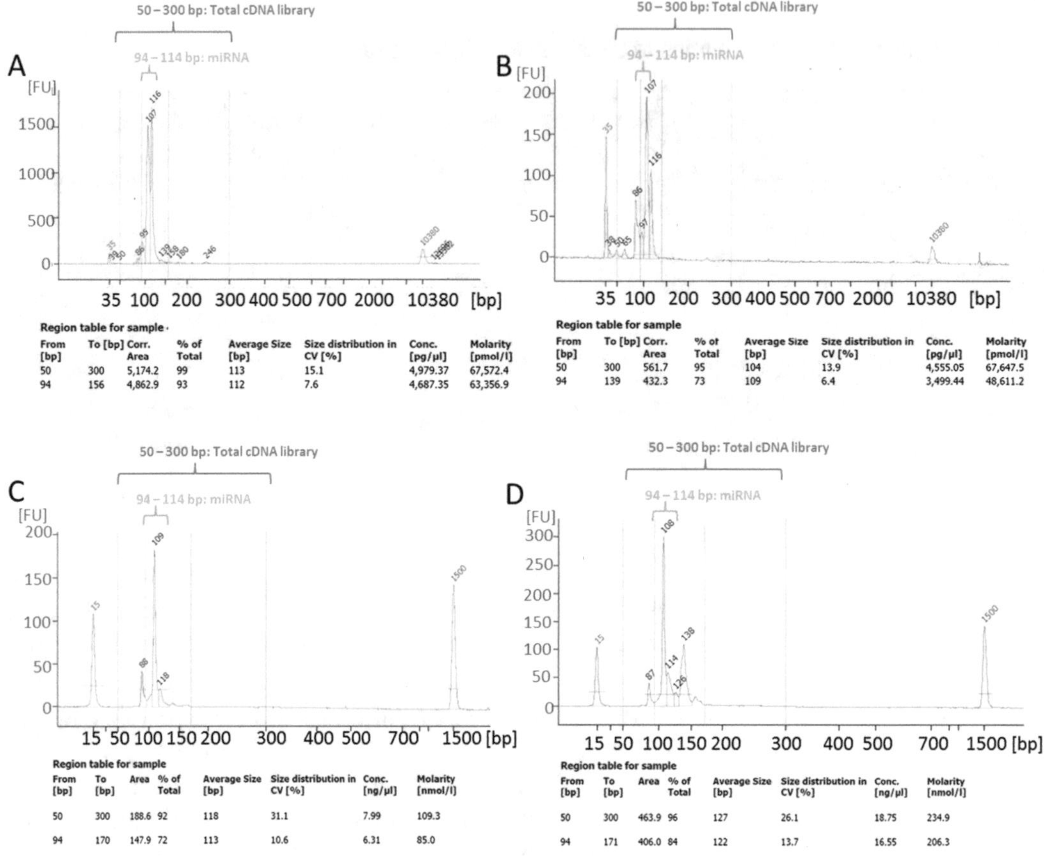

A — Region table for sample

From [bp]	To [bp]	Corr. Area	% of Total	Average Size [bp]	Size distribution in CV [%]	Conc. [pg/µl]	Molarity [pmol/l]
50	300	5,174.2	99	113	15.1	4,979.37	67,572.4
94	156	4,862.9	93	112	7.6	4,687.35	63,356.9

B — Region table for sample

From [bp]	To [bp]	Corr. Area	% of Total	Average Size [bp]	Size distribution in CV [%]	Conc. [pg/µl]	Molarity [pmol/l]
50	300	561.7	95	104	13.9	4,555.05	67,647.5
94	139	432.3	73	109	6.4	3,499.44	48,611.2

C — Region table for sample

From [bp]	To [bp]	Area	% of Total	Average Size [bp]	Size distribution in CV [%]	Conc. [ng/µl]	Molarity [nmol/l]
50	300	188.6	92	118	31.1	7.99	109.3
94	170	147.9	72	113	10.6	6.31	85.0

D — Region table for sample

From [bp]	To [bp]	Area	% of Total	Average Size [bp]	Size distribution in CV [%]	Conc. [ng/µl]	Molarity [nmol/l]
50	300	463.9	96	127	26.1	18.75	234.9
94	171	406.0	84	122	13.7	16.55	206.3

Fig. 2 Examples of small RNA cDNA libraries. The barcoded cDNA library was assessed using a High Sensitivity DNA assay chip (**a** and **b**) or DNA 1000 assay chip (**c** and **d**) and run on the Agilent® Bioanalyzer® instrument. A smear test was performed by gating the size range for all of the ligation products (50–300 bp indicated in *blue*) followed by gating the size range of miRNA (94–114 bp indicated in *green*). For example in **a**, the percentage of ligated miRNA products was 85 % (not shown). However, the size range to include all desired library peaks (94–156 nt) except for cDNA fragments with no insert (<94 nt) was preferred for this particular sample. It was observed that 93 % of the library contained the desired peaks as observed in the region table. The total concentration of the library to be used for template dilution is 67. 575 pM. Sample (**a**) has two prominent peaks: one at 107 bp possibly representing miRNA ligated products and another peak at 116 bp (unknown RNA species) of approximately equal concentrations. (**b**) Contains a larger peak at 107 bp indicating that this sample potentially has more miRNA species than the unidentified peak at 116 bp. (**c**) is highly enriched with miRNA ligated products which displays a peak at 109 bp. Although has little traces of ligated fragments larger than 109 bp. (**d**) is a library which has a prominent peak of miRNA ligated fragments and other small RNA species. Peaks observed at approximately 87 bp are representative of adapter dimers with no insert

2. It is essential that TRIzol® LS is used if exosomes are resuspended in a buffer such as PBS. TRIzol® LS is a concentrated version of TRIzol® and ideal for samples resuspended in buffer. TRIzol® should only be used for cell pellets or tissue samples which are not suspended in buffer.

3. Exosomes resuspended in 250 μl of PBS must maintain the correct sample to TRIzol® LS and chloroform ratio to yield maximal RNA and quality. Furthermore, the aqueous layer from the RNA extraction must be accurately measured in order to determine the correct volume of 100% ethanol to precipitate RNA. The 1.5 volumes of ethanol ensure precipitation of small RNAs and miRNAs.
4. Upon elution of the exosomal RNA from the RNeasy mini column or any other column used, allow the elution buffer to incubate for 1 min to provide RNA sufficient time to unbind from the filter. Furthermore, instead of eluting twice, elute with a larger volume (100 μl), and concentrate the sample with a SpeedVac® before analyzing the RNA profile on a small RNA assay chip.
5. If the yield of small RNA is under 5 ng, concentrate the sample to 3 μl, and use the whole sample for library construction.
6. If you have stored your RNA samples at −20 or −80 °C, ensure you resuspend the RNA samples thoroughly, and briefly centrifuge before transferring the RNA for library construction.
7. For best results, use a new bottle of 100% ethanol for size selection and cleanup procedures. This is to ensure the ethanol has not evaporated. The success of size selection heavily replies on precipitating the correct size range of nucleic acid fragments at a given ethanol concentration. Pre-wetting the pipette tip ensures accurate pipetting of ethanol. Ensure your pipettes are calibrated.
8. If there is not enough cDNA for sequencing, it is possible to add more cDNA into the PCR reaction or increase the number of PCR cycles to 15–20 PCR cycles. However, this must be performed for all samples within the project otherwise batch bias will be observed.
9. If you obtain a low yield of cDNA library, your RNA input may be too low. Optimize your exosome isolation and RNA extraction to ensure the highest miRNA yield possible. Ensure you have greater than 1 ng of miRNA and that the percentage of miRNA in your sample is greater than 1%.
10. Upon diluting the library for template preparation, it is critical to resuspend the library thoroughly once thawed and centrifuge briefly. Dilute the library using low TE buffer and appropriate serial dilutions that enable accurate pipetting and pooling. Vortex each dilution well followed by a brief centrifuge. Do not reuse diluted libraries.

References

1. Bartel DP (2004) MicroRNAs: genomics, biogenesis, mechanism, and function. Cell 116:281–297
2. Griffiths-Jones S (2004) The microRNA registry. Nucleic Acids Res 32:D109–111
3. Cheng L, Sharples RA, Scicluna BJ, Hill AF (2014) Exosomes provide a protective and enriched source of miRNA for biomarker profiling compared to intracellular and cell-free blood. J Extracell Vesicles 3:PMCID:PMC3968297
4. Cheng L, Sun X, Scicluna BJ, Coleman BM, Hill AF (2013) Characterization and deep sequencing analysis of exosomal and non-exosomal miRNA in human urine. Kidney Int 86(2):433–444
5. Andreasen N, Blennow K (2005) CSF biomarkers for mild cognitive impairment and early Alzheimer's disease. Clin Neurol Neurosurg 107:165–173
6. Murata K, Yoshitomi H, Tanida S, Ishikawa M, Nishitani K, Ito H, Nakamura T (2010) Plasma and synovial fluid microRNAs as potential biomarkers of rheumatoid arthritis and osteoarthritis. Arthrit Res Ther 12:R86
7. Valadi H, Ekstrom K, Bossios A, Sjostrand M, Lee JJ, Lotvall JO (2007) Exosome-mediated transfer of mRNAs and microRNAs is a novel mechanism of genetic exchange between cells. Nat Cell Biol 9:654–659
8. Coleman BM, Hanssen E, Lawson VA, Hill AF (2012) Prion-infected cells regulate the release of exosomes with distinct ultrastructural features. FASEB J 26:4160–4173
9. Lässer C (2013) In: Kosaka N (ed) Circulating MicroRNAs:Circulating MicroRNAs: Methods and Protocols. Identification and Analysis of Circulating Exosomal microRNA in Human Body Fluids. Humana Press, Totowa, NJ, Vol 1024, pp. 109–128.
10. Gonzales P, Zhou H, Pisitkun T, Wang N, Star R, Knepper M, Yuen PT (2010) Isolation and purification of exosomes in urine. Methods in molecular biology (Clifton, N.J.), 641, 89–99.
11. Rani S, O'Brien K, Kelleher F, Corcoran C, Germano S, Radomski M, Crown J, O'Driscoll L (2011) Isolation of exosomes for subsequent mRNA, MicroRNA, and protein profiling. Methods in molecular biology (Clifton, N.J.), 784, 181–195.

Chapter 7

A Protocol for Isolation and Proteomic Characterization of Distinct Extracellular Vesicle Subtypes by Sequential Centrifugal Ultrafiltration

Rong Xu, Richard J. Simpson, and David W. Greening

Abstract

Scientific and clinical interest in extracellular vesicles (EVs) has increased rapidly as evidence mounts that they may constitute a new signaling paradigm. Recent studies have highlighted EVs carry preassembled complex biological information that elicit pleiotropic responses in target cells. It is well recognized that cells secrete essentially two EV subtypes that can be partially separated by differential centrifugation (DC): the larger size class (referred to as "microvesicles" or "shed microvesicles," sMVs) is heterogeneous (100–1500 nm), while the smaller size class (referred to as "exosomes") is relatively homogeneous in size (50–150 nm). A key issue hindering progress in understanding underlying mechanisms of EV subtype biogenesis and cargo selectivity has been the technical challenge of isolating homogeneous EV subpopulations suitable for molecular analysis. In this protocol we reveal a novel method for the isolation, purification, and characterization of distinct EV subtypes: exosomes and sMVs. This method, based on sequential centrifugal ultrafiltration (SCUF), affords unbiased isolation of EVs from conditioned medium from a human colon cancer cell model. For both EV subtypes, this protocol details extensive purification and characterization based on dynamic light scattering, cryoelectron microscopy, quantitation, immunoblotting, and comparative label-free proteome profiling. This analytical SCUF method developed is potentially scalable using tangential flow filtration and provides a solid foundation for future in-depth functional studies of EV subtypes from diverse cell types.

Key words Extracellular vesicles, Exosomes, Shed microvesicles, Purification, Sequential centrifugal ultrafiltration

1 Introduction

Extracellular vesicles (EVs) are important mediators of intercellular communication [1]. EVs mediate local and systemic cell communication through the horizontal transfer of information (i.e., mRNAs, microRNAs, and proteins) [2–14]. Scientific and clinical interest in EVs has increased rapidly as evidence mounts that they may constitute a new signaling paradigm [15]. It is clear that EVs carry preassembled complex biological information that elicit pleiotropic responses in various target cells [16–22]. It is well

Andrew F. Hill (ed.), *Exosomes and Microvesicles: Methods and Protocols*, Methods in Molecular Biology, vol. 1545, DOI 10.1007/978-1-4939-6728-5_7,

recognized that cells secrete essentially two EV subtypes that can be readily separated by differential centrifugation: the larger size class (referred to as "microvesicles" or "shed microvesicles," sMVs) is heterogeneous (100–1500 nm) [18, 23–28], while the smaller size class (referred to as "exosomes") is relatively homogeneous in size (50–150 nm) [28, 29]. A key issue hindering progress in understanding underlying mechanisms of EV biogenesis and cargo selectivity has been the technical challenge of isolating homogeneous EV subpopulations suitable for molecular analysis [28, 29].

Effective methods for the isolation and characterization of EVs remain challenging [16, 30–32]. Further, EVs are a heterogeneous group, with the nomenclature still being defined and refined by the research community [33–35]. Current strategies to isolate EVs include differential centrifugation (DC) [36], filtration using hydrophilic polyvinylidene difluoride (PVDF) membranes of different pore sizes [37, 38], high-performance size-exclusion chromatography (SEC) [39], ultrafiltration with SEC [32], immunocapture [18, 40], heparin affinity purification [41], differential density-gradient ultracentrifugation [16], tangential flow filtration [42], field-flow fractionation [43], synthetic polymer-based precipitation [44–46], and microfluidic isolation [47]. Our group recently performed a proteomic analysis evaluating the ability of different techniques (namely, differential ultracentrifugation, OptiPrep density-gradient centrifugation, and immunocapture using EpCAM (CD326) antibodies coupled to magnetic beads to enrich for exosome markers and proteins involved in exosome biogenesis, trafficking, and release). A detailed description [16] and protocol [48] describing these different methods for EV isolation are described. With these improved methods however comes the need to define homogeneous subtypes of EVs, to ensure that subsequent downstream targeted functional analyses to investigate their biological roles are undertaken.

In this protocol we reveal a novel method for the isolation, purification, and characterization of distinct EV subtypes: exosomes and sMVs (Fig. 1). This method, based on sequential centrifugal ultrafiltration (SCUF) of different hydrophilic polyvinylidene difluoride (PVDF) membranes, affords unbiased isolation of EVs from a human colon cancer cell model [28]. For both EV subtypes, this protocol details extensive purification (Fig. 1) and characterization based on dynamic light scattering (Fig. 2), cryo-electron microscopy (Fig. 2), quantitation (Fig. 1), immunoblotting (Figs. 2 and 4), and label-free proteomic profiling (Figs. 3 and 4). Using proteomic-based characterization, we identify 350 proteins that are selectively enriched in sMVs (in comparison with exosomes), many of which have not been previously described in EVs; we expect that many of these identifications will form the basis for definitive sMV protein markers that will enable their distinction from exosomes. Further, we compare this SCUF approach with conventional differential centrifugation approaches in EV

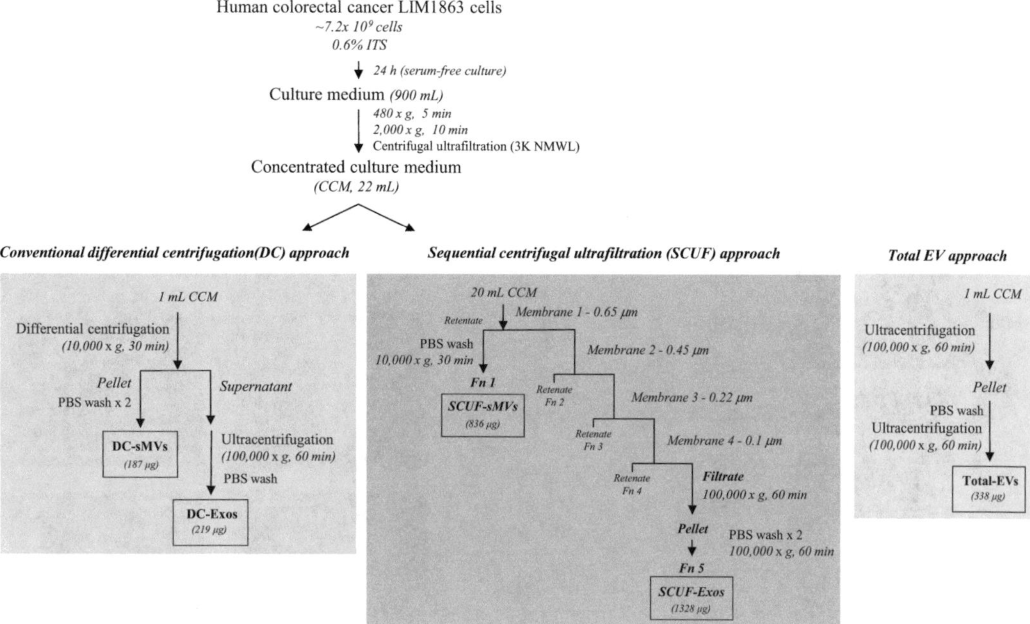

Fig. 1 Isolation of exosomes and shed microvesicles from culture medium. Human cancer cells (LIM1863) were grown for 24 h in serum-free culture conditions, with culture medium isolated, and concentrated (CCM). Several different EV isolation methods were employed including (**a**) conventional differential centrifugation (DC) approach, (**b**) sequential centrifugal ultrafiltration (SCUF) approach, and (**c**) total EV approach. For conventional DC approach, CCM was centrifuged at 10,000 × *g* for 30 min (to isolate DC-sMVs, 187 μg) and the supernatant sequentially ultracentrifuged at 100,000 × *g* for 1 h (to isolate DC-Exos, 219 μg). For the SCUF approach, CCM (20 mL) was fractionated using a combination of different molecular pore-sized ultrafilters (0.65–0.1 μm). SCUF-sMVs were isolated using 0.65 μm membrane filter (Fn 1, 836 μg). Following sequential ultrafiltration of the <0.65 μm filtrate, SCUF-Exos were isolated using a 0.1 μm membrane filter (Fn 5, 1328 μg). For total EVs (Exos and sMVs), CCM (1 mL) was ultracentrifuged at 100,000 × *g* for 30 min, with PBS wash (total EVs, 338 μg). The protocol describes each of the different methods for EV isolation

isolation and characterization (Fig. 1), indicating the selectivity of this methodology. This analytical workflow is potentially scalable using tangential flow filtration and provides a solid foundation for future in-depth functional studies of EV subtypes from diverse cell types and an increased range of functional assays.

2 Materials

2.1 Cell Culture and Concentrated Culture Medium (CCM) Preparation

1. Human LIM1863 colorectal cancer cells [49] (*see* **Note 1**).
2. RPMI-1640 media (#11875119, Life Technologies).
3. Tissue culture flasks T150 cm^2 (#355001, BD Falcon).
4. *Cell culture medium*: 5% (v/v) fetal calf serum (FCS), 0.1% (v/v) Insulin-Transferrin-Selenium (ITS) (#51300-044, Life Technologies), and 60 μg/mL benzylpenicillin and 100 μg/mL

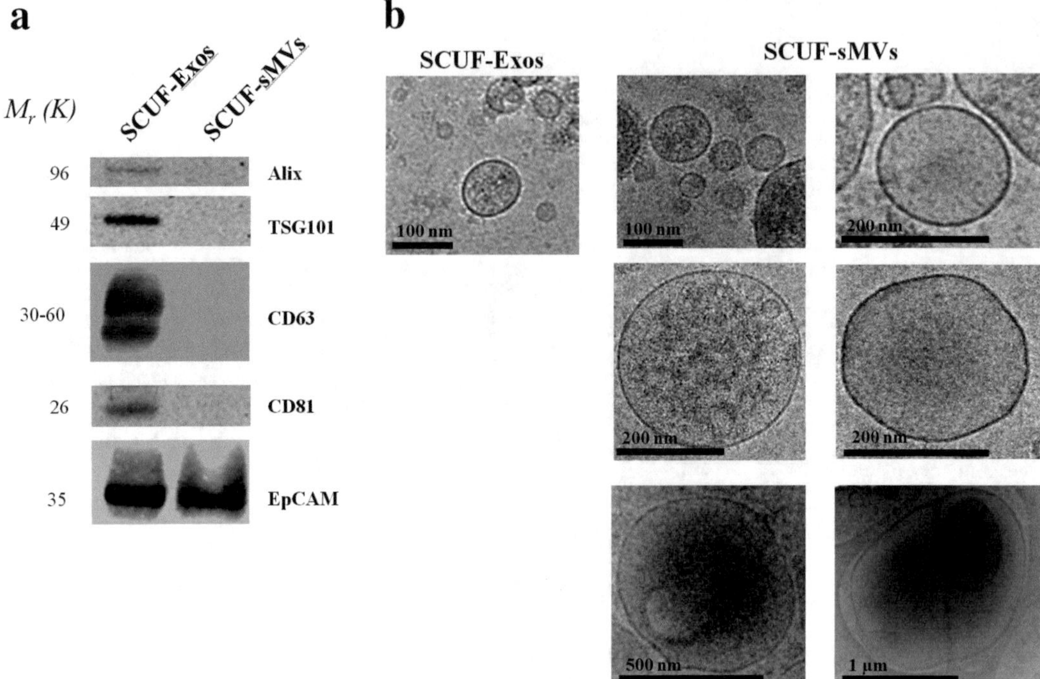

Fig. 2 Characterization of distinct exosomes and shed microvesicles isolated by SCUF. (**a**) For Western blotting, SCUF-Exos and SCUF-sMVs were probed with anti-mAbs to Alix, TSG101, CD63, CD81 and EpCAM. (**b**) Size and morphology of SCUF-Exos and SCUF-sMVs were visualized by cryo-EM. Scale bar shown for individual size ranges. (**c**) Size distribution of SCUF-Exos and SCUF-sMVs (measured from ~200 cryoelectron micrographs from five fields of view. (**d**) Dynamic light scattering analysis of CCM, SCUF-Exos and SCUF-sMVs. Mean hydrodynamic diameter of EVs was calculated by fitting a Gaussian function to the measured size distribution

streptomycin (P/S) (#15140-122, Life Technologies) supplemented in RPMI-1640 media.

5. *Serum-free cell culture medium*: 0.6% (v/v) ITS and P/S in RPMI-1640 media [48] (*see* **Note 2**).
6. 50 mL polypropylene centrifuge tubes (three separate 50 mL tubes required for every 50 mL of CM collected due to sample processing) (#352070, Falcon tubes, Life Technologies).
7. Refrigerated centrifuge (to 4 °C), centrifugation capability to 3000 × *g*.
8. Polyallomer tubes, appropriate for the ultracentrifuge rotor.
9. Amicon Ultra-15, Ultracel 3K centrifugal filter devices (#UFC900308, Merck Millipore).

2.2 EV Isolation (See Notes 3 and 4)

2.2.1 Conventional Differential Centrifugation (DC) Approach

1. CCM (1 mL 1.9 mg protein).
2. Ultracentrifuge and matched rotor (TLA-55) (Optima MAX-XP, Beckman Coulter).
3. Sterile/filtered PBS.

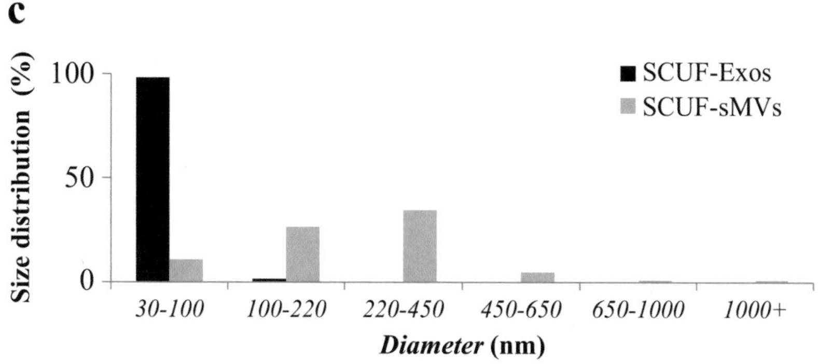

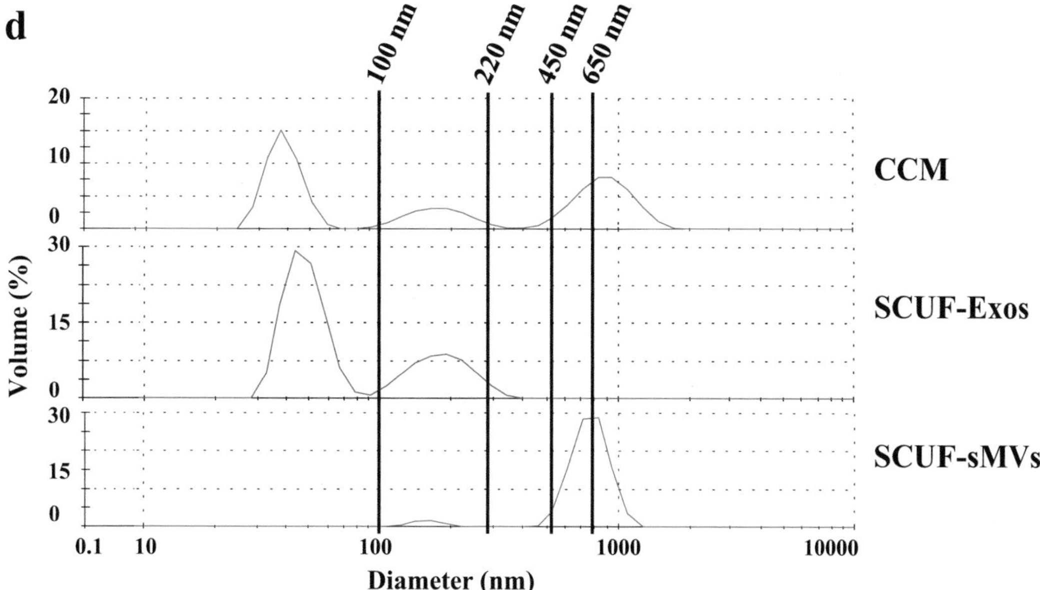

Fig. 2 (continued)

4. TLA-55 fixed angle (small scale/washing)—Microcentrifuge Polypropylene Tube (#357448, Beckman Coulter).

2.2.2 Sequential Centrifugal Ultrafiltration (SCUF) Approach (See Note 5)

1. CCM (20 mL, 38 mg protein).
2. Different pore-sized PVDF ultrafilters (depending on range selectivity). For this protocol we utilize 0.65, 0.45, 0.22, and 0.1 μm filter membrane (#UFC40VV25, 0.65 μm; UFC40GV25, 0.22 μm; UFC40HV25, 0.45 μm; UFC40DV25, 0.65 μm; Durapore Ultrafree-CL, Merck Millipore).
3. Refrigerated centrifuge (to 4 °C), centrifugation capability to 3000 × *g*.
4. TLA-55 fixed angle (small scale/washing)—Microcentrifuge Polypropylene Tube (#357448, Beckman Coulter).

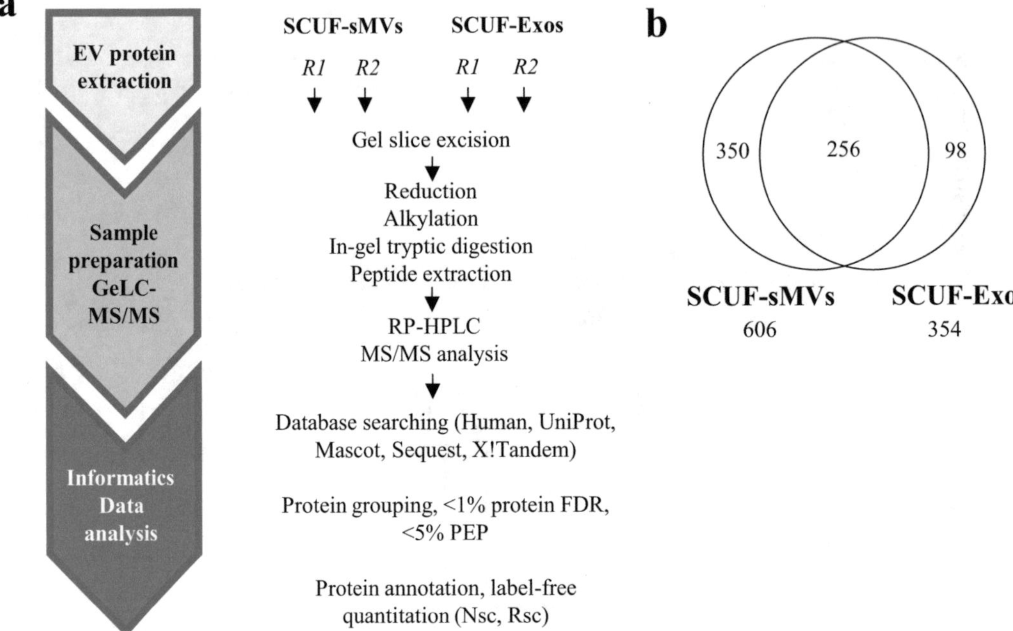

Fig. 3 Proteome analysis of exosomes and shed microvesicles using SCUF. (**a**) Proteomic profiling of the two distinct EV subtypes, SCUF-sMVs and SCUF-Exos, by the workflow shown. Stringent data validation and label-free quantitation based on normalized spectral count (Nsc) ratios was performed. (**b**) The distribution of identified proteins is shown in a two-way Venn diagram of SCUF-sMVs and SCUF-Exos with 256 proteins commonly identified, and 350 and 98 proteins uniquely identified in SCUF-sMVs and SCUF-Exos, respectively. (**c**) For each individual protein, significant peptide MS/MS spectra were summated and normalized by the total number of significant peptide MS/MS spectra identified in the sample. The ratio serves an indicator of protein abundance, i.e., the higher the ratio, the more abundant the protein within the sample. Protein categories of interest included (1) exosome markers, (2) cytoskeleton network components, and (3) cargo trafficking and sorting proteins

5. Ultracentrifuge (Optima MAX-XP, Beckman Coulter).
6. Sterile/filtered PBS.

2.2.3 Total EV Isolation

1. CCM (1 mL, 1.9 mg protein).
2. Ultracentrifuge and matched rotor (TLA-55) (Optima MAX-XP, Beckman Coulter).
3. Sterile/filtered PBS.
4. TLA-55 fixed angle (small scale/washing)—Microcentrifuge Polypropylene Tube (#357448, Beckman Coulter).

2.3 EV Characterization (See Note 6)

2.3.1 Buoyant Density Analysis (OptiPrep 5–40 % Gradient) (See Note 7)

1. EVs prepared by subheading 2.2.1 as indicated (DC approach) in 500 μL PBS.
2. SW40 Ti rotor-Thinwall Polypropylene tubes (#345775, Beckman Coulter).
3. Ultracentrifuge (Optima TM XPN, Beckman Coulter).

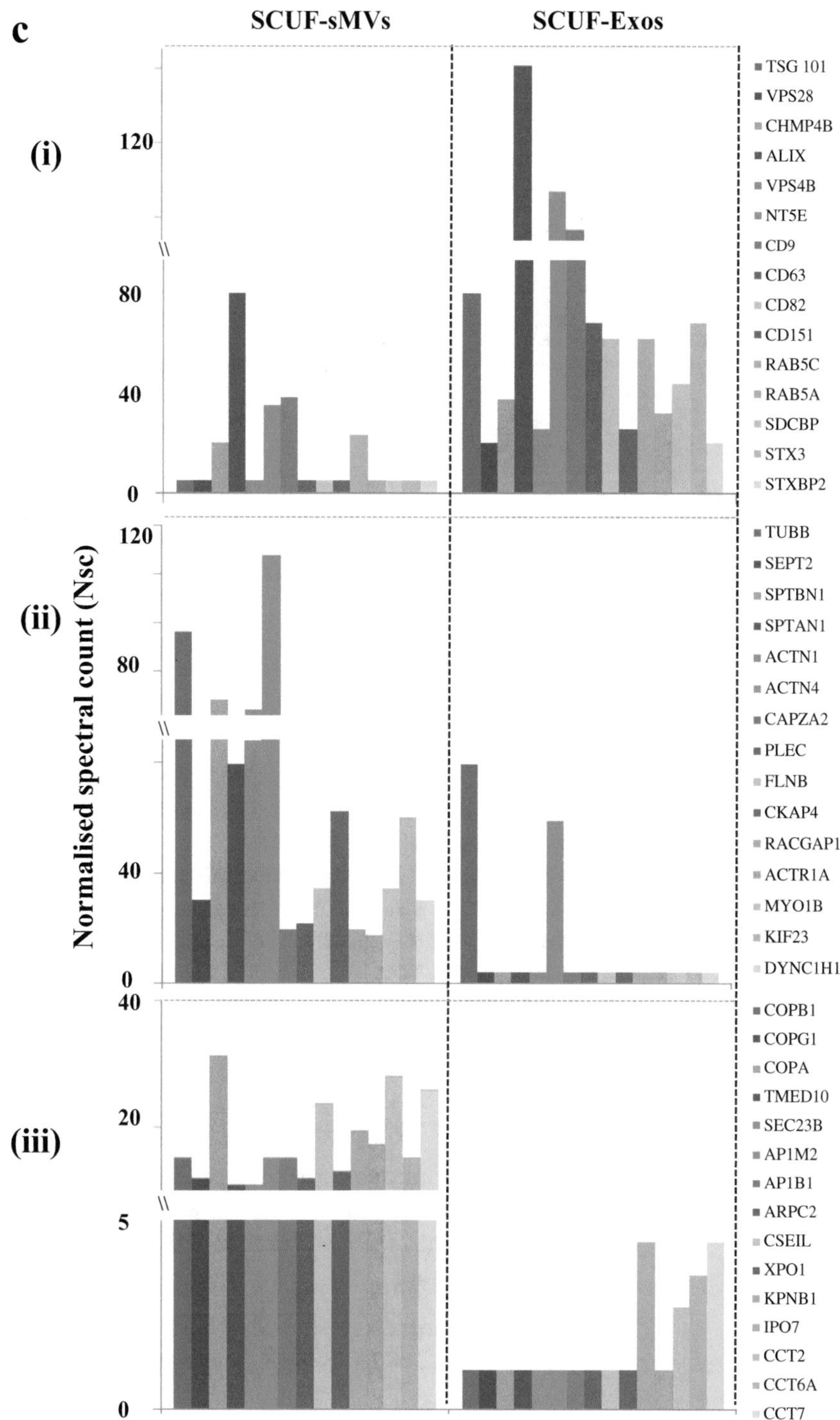

Fig. 3 (continued)

a

	Protein Acc	Gene Name	Protein Description	SCUF-Exos[a]	SCUF-sMV[b]	Rsc (SCUF-Exos v SCUF-sMV)[c]
sMV marker	Q02241	KIF23	Kinesin-like protein KIF23	0	27	-14.1
	P55060	XPO/CSE1L	Exportin-2	0	21	-11.1
Exosome marker	Q99816	TSG101	Tumor susceptibility gene 101	12	0	17.0

a - Significant MS/MS spectral counts identified in SCUF-Exos
b - Significant MS/MS spectral counts identified in SCUF-sMVs
c - Relative spectral count ratio (Rsc) for proteins identified in SCUF-sMVs, compared with SCUF-Exos

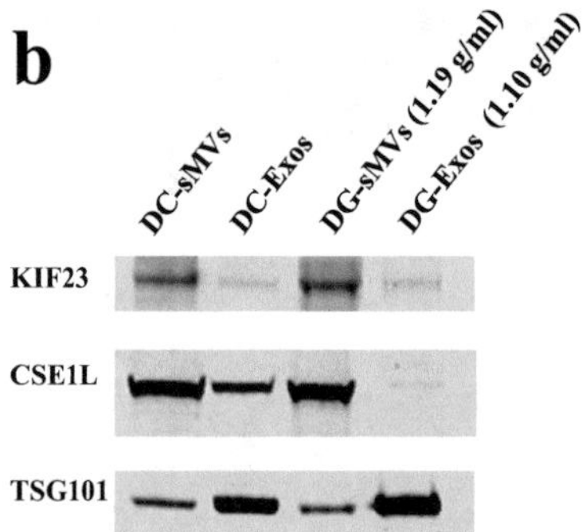

Fig. 4 Validation of sMV markers: KIF23 and CSE1L. (**a**) For SCUF-sMVs and SCUF-Exos, relative spectral count ratios (Rsc) were determined as detailed in Eq. 2. Unique expression of KIF23 and CSE1L were defined as sMV markers, while TSG101 was confirmed as exosomal marker for comparison. Rsc serves an indicator of protein abundance, i.e., the higher the ratio, the more abundant the protein within the sample. (**b**) To confirm expression of sMV and exosomal markers, Western blotting was performed (15 μg protein) for several EV preparations. These included DC-sMVs, DC-Exos, OptiPrep density-gradient separated (DG)-sMVs (1.19 g/mL fraction) and DG-Exos (1.10 g/mL fraction). Fractions were probed with anti-mABs directed to KIF23, CSE1L, and TSG101, demonstrating the selectivity of sMV and exosomal marker proteins

4. Stock solution of OptiPrep (60% (w/v) aqueous iodixanol (Axis-Shield PoC, Norway).
5. Prepare discontinuous iodixanol gradient, 40% (w/v), 20% (w/v), 10% (w/v), and 5% (w/v) solutions of iodixanol by diluting stock with 0.25 M sucrose/10 mM Tris, pH 7.5.
6. Sterile/filtered PBS.
7. TLA-55 fixed angle (small scale/washing)—Microcentrifuge Polypropylene Tube (#357448, Beckman Coulter).

2.3.2 Protein Quantitation (Densitometry-Based) (See Note 8)

1. SDS sample buffer (4% (w/v) sodium dodecyl sulfate, 125 mM Tris-HCl, pH 6.8, 20% (v/v) glycerol, 0.01% bromophenol blue, 50 mM dithiothreitol (DTT)). This can be prepared in bulk, aliquoted, and stored at −20 °C.
2. CCM from preparations in Subheading 2.1.

3. NuPAGE 1-mm 12-well 4–12% (w/v) Bis-Tris Precast gels (Life Technologies).
4. NuPAGE 1 × MES running buffer (#NP0002-02, Life Technologies).
5. XCell SureLock gel tank (Life Technologies).
6. BenchMark Protein Ladder standard of known protein concentration (1.7 μg/μL) (#10747-012, Life Technologies).
7. SYPRO Ruby fluorescent stain for protein detection separated by polyacrylamide gel electrophoresis (PAGE) (#S-21900, Life Technologies).
8. SYPRO Ruby fix solution (40% (v/v) methanol, 10% (v/v) acetic acid in water).
9. SYPRO Ruby destaining solution (10% (v/v) methanol with 6% (v/v) acetic acid).
10. Orbital shaker.
11. Typhoon 9410 variable mode imager, green (532 nm) excitation laser, and 610BP30 emission filter (Molecular Dynamics).
12. ImageQuant software (Molecular Dynamics) or suitable densitometry-based analysis software.

2.3.3 Western Blot Analysis (See Note 6)

1. iBlot Dry Blotting System and transfer membranes (#IB1001, Invitrogen).
2. 5% (w/v) skim milk powder in Tris-buffered saline (50 mM Tris, 150 mM NaCl) with 0.05% (v/v) Tween-20 (TTBS).
3. Mouse anti-TSG101 (BD Biosciences; 1:500) in TTBS.
4. Mouse anti-CD81 (Santa Cruz Biotechnology; 1:1000) in TTBS.
5. Mouse anti-Alix (Cell Signaling Technology; 1:1000) in TTBS.
6. Mouse anti-KIF23 (Santa Cruz Biotechnology; 1:1000) in TTBS.
7. Mouse anti-CSE1L (Santa Cruz Biotechnology; 1:1000) in TTBS.
8. Mouse anti-EPCAM (nonreducing) (Santa Cruz Biotechnology; 1:1000) in TTBS.
9. Mouse anti-CD63 (nonreducing) (Santa Cruz Biotechnology; 1:1000) in TTBS.
10. IRDye 800 goat anti-mouse IgG (1:15,000, LI-COR Biosciences).
11. Orbital shaker.
12. Odyssey Infrared Imaging System, v3.0 (LI-COR Biosciences, Nebraska, USA).

2.3.4 Dynamic Light Scattering (DLS) (See Note 9)

1. CCM (50 μL), EV preparations for DLS (~10 μg proteins in 50 μL PBS).
2. Sterile/filtered PBS.
3. Titertek shaker (Flow Laboratories, Inc.).
4. Disposable micro cuvettes (ZEN0040, Malvern Instruments Ltd., UK).
5. Zetasizer Nano ZS (Malvern Instruments Ltd., UK).

2.3.5 Cryoelectron Microscopy (cryo-EM) (See Note 10)

1. EV preparations (~2 μg protein).
2. Aurion Protein-G gold 10 nm (ProSciTech, QLD, Australia).
3. Sterile/filtered PBS.
4. Glow-discharged C-flat holey carbon grids (ProSciTech).
5. Liquid ethane.
6. Liquid nitrogen.
7. Gatan cryoholder (Gatan, Inc., Warrendale, PA, USA).
8. Tecnai G2 F30 (FEI, Eidhoven, NL).

2.3.6 Nanoparticle Tracking Analysis (NTA) (See Note 11)

1. CCM (1–2 μL) and EV preparations (1–2 μg/μL).
2. NanoSight NS300 equipped with 405 nm (violet) laser, sCMOS camera, and NTA software 3.1 build 3.1.45 (NanoSight Ltd., Minton Park, UK).
3. 1 mL disposable syringe.
4. Sterile/filtered PBS.

2.4 Proteomic Sample Preparation and Analysis (See Note 12)

1. Imperial Protein Stain (#24615, Pierce, Thermo Fisher Scientific).
2. Protein LoBind Tubes—1.5 mL microcentrifuge tube (low protein binding, Eppendorf, #022431081) or Protein LoBind Plates (low protein binding, Eppendorf, #951032905).
3. 100 mM ammonium bicarbonate (NH_4HCO_3: 0.4 g in 50 mL water).
4. 50 mM ammonium bicarbonate/acetonitrile (1:1 v/v).
5. 50 mM ammonium bicarbonate in water.
6. 10 mM dithiothreitol (DTT) (Calbiochem, San Diego, USA) in 100 mM ammonium bicarbonate (7.5 mg DTT).
7. 50 mM iodoacetamide (IAA) (Fluka, St. Louis, USA) in 100 mM ammonium bicarbonate (10 mg IAA).
8. 1 mL trypsin buffer: 10 mM ammonium bicarbonate, 10% acetonitrile.
9. Trypsin stock solution: dissolve content of a 20 μg vial (V5111, 5×20 μg, Promega) in 20 μL of trypsin buffer (10 mM

ammonium bicarbonate, 10% acetonitrile) and keep on ice. The concentration of trypsin is 1 μg/μL.

10. 10 mL of 5% TFA in water (v/v).
11. Buffer A: 0.1% TFA, 2% acetonitrile in water.
12. 2 mL Extraction Buffer: (A) 30% acetonitrile in 0.1% TFA and water; (B) 50% acetonitrile in 0.1% TFA and water; (C) 85% acetonitrile in 0.1% TFA and water.
13. Thermomixer temperature range up to 56 °C.
14. Thermostat oven at 37 °C.
15. Sonicator.
16. Vacuum centrifuge (lyophilizer).
17. STAGE-Tip/desalting column—remove small disks (2–3) of C18 Empore filter using a 22 G flat-tipped syringe and ejecting disks into P200 pipette tips or a commercial one (#SP301 STAGE-Tips, C18 material, 200 μL tip, Thermo Fisher Scientific). Ensure that the disk is securely wedged in the bottom of the tip. Condition the columns for each sample (utilize Extraction Buffer).
18. MS sample vials with snap lid (#THC11141190, Snap ring vial with glass insert, Thermo Fisher Scientific).
19. Nanoflow UPLC instrument (Ultimate 3000 RSLCnano Thermo Fisher Scientific).
20. Orbitrap Elite mass spectrometer (Thermo Fisher Scientific) equipped with a nanoelectrospray ion source (Thermo Fisher Scientific).
21. Pre-column: Acclaim PepMap100 C18 5 μm 100 Å (Thermo Fisher Scientific).
22. VYDACMS C18-reversed phase column (25 cm length, 75 μm inner diameter, 3 μm 300 Å) (Grace, Hesperia, CA).
23. Xcalibur software v2.1 (Thermo Fisher Scientific).
24. Proteome Discoverer (v1.4.0.288, Thermo Fisher Scientific).
25. Mascot (Matrix Science, London, UK; v 1.4.0.288).
26. Sequest (Thermo Fisher Scientific, San Jose, CA, v 1.4.0.288).
27. X! Tandem (v 2010.12.01.1).
28. Scaffold (Proteome Software Inc., Portland, OR, v 4.3.4).

3 Methods

In this protocol we reveal a novel method for the isolation, purification, and characterization of distinct EV subtypes: exosomes and sMVs (Fig. 1). This method, based on sequential centrifugal ultrafiltration (SCUF) of different hydrophilic polyvinylidene difluoride (PVDF) membranes, affords unbiased isolation of EVs from a human colon cancer cell model. For these distinct EV subtypes, this protocol details extensive purification (Fig. 1) and characterization based on dynamic light scattering (Fig. 2), cryoelectron microscopy (Fig. 2), quantitation (Fig. 1), immunoblotting (Figs. 2 and 4), and label-free proteomic profiling (Figs. 3 and 4). Further, we compare this SCUF approach with conventional differential centrifugation approaches in EV isolation and characterization (Fig. 1), identify several new sMV protein markers, and indicate the selectivity of the SCUF methodology.

3.1 Cell Culture and CCM Preparation

1. As a model, human colon carcinoma LIM1863 cell organoids [49] were cultured in T150 cm^2 flasks in (300 mL) cell culture medium (5% FCS, 0.1% ITS) at 10% CO_2, 37 °C (*see* **Note 1**).
2. Cells (~7.2 × 10^9) were washed four times with 30 mL of serum-free media (or sterile PBS) and cultured for 24 h in 900 mL serum-free cell culture medium (*see* **Note 2**).
3. Culture medium (CM) is collected (~900 mL) into 50 mL polypropylene tubes and centrifuged at 4 °C (480 × *g* for 5 min, 2000 × *g* for 10 min) to remove intact cells and cell debris. The CM supernatant is retained and contains both soluble (cytokines, growth factors, etc.) and membrane vesicle components.
4. CM is concentrated using Amicon Ultra-15, Ultracel centrifugal filter devices with 3K NMWL (3000 × *g*). Several centrifugal filter devices can be used for the same sample to increase efficiently of concentration, before combining retentate into one centrifugal filter for further concentration (95% of filtrate filtered).
5. CCM storage: Short-term on ice (within 3 days), long-term up to 6 months −80 °C.

3.2 EV Isolation (See Notes 3 and 4)

3.2.1 Conventional Differential Centrifugation (DC) Approach

1. For isolation of sMVs, 1 mL of CCM is centrifuged at 10,000 × *g* for 30 min to isolate sMVs. The pellet is resuspended in 1 mL of sterile/filtered PBS, before further centrifugation at 10,000 × *g* for 30 min to obtain the washed sMV pellet (referred to as DC-sMVs). sMVs are either used immediately (on ice) or stored at −80 °C.
2. To isolate Exos, the supernatant after the initial 10,000 × *g* centrifugation step is isolated and ultracentrifuged at 100,000 × *g* for 60 min at 4 °C. The pellet is resuspended in 1 mL sterile/filtered PBS and re-centrifuged (100,000 × *g*, 60 min), resuspended in 50 μL PBS to obtain washed exosomes (referred to as DC-Exos). Exosomes are either used immediately (on ice) or stored at −80 °C.

3.2.2 Sequential Centrifugal Ultrafiltration (SCUF) Approach (See Note 5)

1. PVDF ultrafilters were washed by 500 μL PBS and centrifuged at 3000 × *g* for 10 s, discard PBS and put on ice.
2. CCM was fractionated using a series of different pore-size PVDF ultrafilters (Durapore Ultrafree-CL) ranging from 0.65, 0.45, 0.22, and 0.1 μm.
3. CCM was initially centrifuged at 3000 × *g* using 0.65 μm ultrafilters (95% of the filtrate retained). Filters were washed with 500 μL PBS. The retentate (SCUF-fraction >0.65 μm, fraction 1 (Fn1)) from this step was transferred into a fresh micro-tube, washed with 0.5 mL PBS, and harvested by centrifugation at 10,000 × *g* for 30 min.
4. The Fn1 filtrate (i.e., <0.65 μm) was sequentially filtered through 0.45 μm (Fn2), 0.22 μm (Fn3), and 0.1 μm (Fn4) filters.
5. All fractions were individually harvested by ultracentrifugation at 100,000 × *g* for 60 min.
6. For the fraction filtered through the 0.1 μm filter (Fn5), this fraction was washed with 0.5 mL PBS and re-centrifuged at 100,000 × *g* for 60 min.
7. All EV preparations (i.e., Fn1–5) were resuspended in 500 μL PBS. Fractions are either used immediately (on ice) or stored at −80 °C.

3.2.3 Total EV Isolation

1. For comparison of DC and SCUF approaches with total EVs, both sMVs and Exos were isolated using ultracentrifugation. For isolation of total EVs, 1 mL of CCM is centrifuged at 100,000 × *g* for 60 min, the pellet resuspended in 1 mL of sterile/filtered PBS, before further centrifugation at 100,000 × *g* for 60 min to obtain the washed total EV pellet (referred to as total EVs). Total EV fraction is either used immediately (on ice) or stored at −80 °C.

3.3 EV Characterization

3.3.1 Buoyant Density Analysis (OptiPrep 5–40% Gradient) (See Note 7)

1. Each EV preparation (DC-Exos, 500 μL, 578 μg proteins; DC-sMVs, 500 μL, 893 μg proteins) overlaid on top of the discontinuous iodixanol gradient (40% (w/v), 20% (w/v), 10% (w/v), and 5% (w/v) solutions of iodixanol was made by diluting a stock solution of OptiPrep with 0.25 M sucrose/10 mM Tris, pH 7.5). The gradient was formed by adding 3 mL of 40% iodixanol solution, followed by careful layering of 3 mL each of 20% and 10% solutions and 2.5 mL of the 5% solution.
2. EV preparation density solution is centrifuged at 100,000 × *g* (TLA-45 fixed angle) for 18 h at 4 °C.
3. Twelve individual 1 mL gradient fractions collected manually for each preparation (with increasing density—top–bottom collection).

4. For both preparations (i.e., DC-Exos and DC-sMVs), fractions were diluted with 2 mL PBS and centrifuged at 100,000 × *g* for 3 h at 4 °C followed by washing with 1 mL PBS and resuspended in 50 μL PBS. DG-sMV buoyant density determined (1.19 g/mL), and DG-Exos buoyant density determined (1.10 g/mL).
5. To determine density of each fraction, a control OptiPrep gradient containing 500 μL of 0.25 M sucrose/10 mM Tris, pH 7.5 was run in parallel. Fractions were collected as described, serially diluted 1:10,000 with water, and the iodixanol concentration determined by absorbance at 244 nm using a molar extinction coefficient of 320 l g^{-1} cm^{-1} [50].

3.3.2 Protein Quantitation (Densitometry-Based) (See Note 8)

1. Samples (CCM or purified EVs) (5 μL) were solubilized in SDS sample buffer and loaded on NuPAGE 4–12% (w/v) Bis-Tris Precast gels.
2. 5 μL BenchMark Protein Ladder (1.7 μg/μL) is loaded into a separate well on NuPAGE 4–12% (w/v) Bis-Tris Precast gels for quantitation.
3. Electrophoresis was performed at 150 V for 1 h in NuPAGE 1× MES running buffer.
4. After power off, the gels were removed from the tank and fixed in 50 mL fixing for 30 min on an orbital shaker and stained with 30 mL SYPRO Ruby for 60 min, followed by destaining for 1 h at RT.
5. Gel is scanned and fluorescent intensity imaged on a Typhoon 9410 variable mode imager, using a green (532 nm) excitation laser and a 610BP30 emission filter at 100 μm resolution.
6. Densitometry quantitation was performed using ImageQuant software to determine protein concentration relative to a BenchMark Protein Ladder standard of known protein concentration (1.7 μg/μL). SYPRO Ruby is a highly sensitive fluorescent stain of proteins to detect linear quantitation range over three orders of magnitude, also being compatibility with mass spectrometry.

3.3.3 Western Blot Analysis (See Note 6)

1. Samples (~15 μg protein) lysed in SDS sample buffer with 50 mM DTT and heated for 5 min at 95 °C (closed cap, although cap can be opened for evaporation to reduce load volume in the gel well).
2. Electrophoresis performed on lysed samples at constant 150 V for 1 h.
3. Proteins electro-transferred onto nitrocellulose membranes using iBlot Dry Blotting System.

4. Membranes blocked with 5 % (w/v) skim milk powder in Tris-buffered saline with 0.05 % (v/v) Tween-20 (TTBS) for 1 h at RT.
5. Membranes probed with primary antibodies (anti-TSG101, anti-EPCAM (no reducing condition), anti-Alix, anti-CD63 (no reducing condition), anti-CD81, anti-KIF23, and anti-CSE1L) for 1 h at RT or overnight at cold room in TTBS followed by incubation with secondary antibody, IRDye 800 goat anti-mouse IgG for 1 h in darkness at RT (Figs. 2 and 4).
6. All antibody incubations were carried out using gentle orbital shaking.
7. Western blots were washed three times in TTBS for 10 min after each incubation step and visualized using the Odyssey Infrared Imaging System.

3.3.4 Dynamic Light Scattering (DLS) (See Note 9)

1. For DLS analyses, 50 μL CCM or 10 μg proteins of each EV preparation in 50 μL PBS were shaken (Titertek shaker, Flow Laboratories, Inc.) at 4 °C for 20 min to dissociate possible EV aggregates.
2. Disposable micro cuvettes (ZEN0040, Malvern Instruments Ltd., UK) were used and DLS measurements conducted at 20 °C using a Zetasizer Nano ZS (Malvern Instruments Ltd., UK), operated at 633 nm and backscattering at 173°. Light scattering was recorded for 200 s with 10 replicate measurements.
3. DLS signal intensity was transformed to volume distribution [volume (%)], assuming a spherical shape of EVs, using the Dispersion Technology Software v.5.10 (Malvern Instruments Ltd., UK) (Fig. 2).

3.3.5 Cryo-transmission Electron Microscopy (cryo-EM) (See Note 10)

1. EV preparations (~2 μg protein) mixed with Aurion Protein-G gold 10 nm at 1:3 ratios. This will allow vesicle diameters to be determined and the possibility for tomographic data collection.
2. Samples transferred onto glow-discharged C-flat holey carbon grids and excess liquid blotted. Grids immediately plunge-frozen in liquid ethane.
3. Grids mounted in a Gatan cryoholder in liquid nitrogen.
4. Images acquired at 300 kV using a Tecnai G2 F30 in low-dose mode.
5. Typically 5–10 fields of view are obtained (Fig. 2).

3.3.6 Nanoparticle Tracking Analysis (NTA) (See Note 11)

1. For NTA analyses, 1–2 μL CCM or 1–2 μg EV preparations were diluted into 1 mL PBS in 1.5 mL tubes.
2. 1 mL disposable syringe was used to inject sample (1 mL) into O-ring top plate in NanoSight NS300 instrument equipped with a 405 nm laser and sCMOS camera.

3. Videos were collected and analyzed using the NTA software 3.1 build 3.1.45. Temperature was recorded at 25 °C. Ensure 20–40 particles are within the field of view and $2–10 \times 10^8$ of typical particle concentration; the image is optimized to identify camera level, sample concentration, beam position, and focus.
4. For each sample, multiple videos ($n = 5$) of typically 60 s duration were taken.
5. Post-acquisition settings were optimized with appropriate detection threshold maintained between samples. Replicate histograms for each sample were averaged with EV size and particle concentration.

3.3.7 *Proteomic Analysis*

Protein Extraction, Reduction, Alkylation (*See Note* 12)

1. EV preparations (10 μg) lysed in SDS sample buffer and heated at 95 °C for 10 min, and proteins separated by SDS-PAGE (approximately 15 mm into the gel) and visualized by Imperial Protein Stain, and gel was destained with ddH_2O.
2. For each band prepare a 1.5 mL microcentrifuge tube with 500 μL of 50 mM ammonium bicarbonate/acetonitrile (1:1 v/v).
3. Each gel lane was excised, and individual gel pieces were transferred carefully into microcentrifuge tubes.
4. For sample reduction, microcentrifuge tubes are centrifuged briefly, heated using a thermomixer at 56 °C (700 rpm for 15 min), and the solution discarded. 200 μL acetonitrile is added, and gel pieces should shrink and take an opaque-white color. Remove acetonitrile and air-dry for 5 min in thermomixer at 56 °C. Add 50 μL fresh DTT solution and incubate at 56 °C for 30 min.
5. For sample alkylation, set thermomixer to 22 °C, and remove DTT solution completely. Immediately add 70 μL IAA solution, and incubate for 20 min in thermomixer 22 °C (700 rpm) covered by aluminum foil.
6. The IAA solution is then removed, and 300 μL acetonitrile added for 2 min and then removed. Add 100 μL of 50 mM ammonium bicarbonate/acetonitrile (1:1 v/v), and incubate for 30 min in thermomixer with light mixing at RT. The sample should be lightly centrifuged and supernatant removed.
7. 100 μL acetonitrile added to sample, removed completely, and air-dried for 5 min.

Tryptic Digestion and STAGE-Tip Desalting

1. For tryptic digestion, add enough trypsin to cover the dry gel pieces (typically, 60–70 μL trypsin solution depending on gel volume with 500 ng trypsin). Store all samples immediately on ice. After 30 min, check if all solution is absorbed and add more trypsin, if necessary. Gel pieces should be completely covered with trypsin.

2. Leave gel pieces for another 30 min to saturate with trypsin, and add 20 μL of 50 mM ammonium bicarbonate to cover the gel pieces.
3. Place tubes with gel pieces into the thermostat oven, and incubate samples overnight (around 16 h) at 37 °C.
4. Add 2 μL of 5 % TFA to each digest, briefly vortex and spin (for samples >80 μL volume then use 4 μL).
5. Withdraw supernatants to low protein tubes; use a pipette with fine gel loader tip (reuse these tips for all following peptide collection steps).
6. Add 70 μL of Extraction Buffer B to each tube, and sonicate for 15 min on the low power setting.
7. Centrifuge briefly, and collect the supernatants into their corresponding low-bind tube by reusing the tip.
8. And repeat **steps 6** and 7 using Extraction Buffer A and Extraction Buffer B, respectively.
9. Add 50 μL 100% acetonitrile, and let the gel pieces to shrink, and take an opaque-white color, no sonication required.
10. Dry down samples (covered with Vacufilm with four pin holes) in vacuum centrifuge for 15 min.
11. Re-acidify with 5 μL of 5% TFA. Check and make sure pH < 2.5.
12. For STAGE-Tip desalting, prepare as many desalting columns as necessary by punching out small disks (2–3) of C18 Empore filter using a 22 G flat-tipped syringe and ejecting the disks into P200 pipette tips. Ensure that the disk is securely wedged in the bottom of the tip.
13. Condition columns by forcing 100% acetonitrile through (25 μL), and check whether the STAGE-Tips are leaky. A benchtop centrifuge of maximum centrifugation force of 400 × *g* should be used to all STAGE-Tip filtering.
14. Remove any remaining organic solvent in the column by forcing Buffer A (0.1% TFA, 2% acetonitrile) through the disk (×2) (30 μL).
15. Force the acidified peptide sample (pH < 2.5) through the C18-STAGE-Tip column, and wash the column with 20 μL buffer A.
16. Transfer STAGE-Tips to new tubes, and elute the peptides from the C18 material using 20–30 μL Extraction Buffer B (×2), 20 μL Extraction Buffer A, and 20 μL Extraction Buffer C.
17. Carefully dry samples in the speed vac without heating, until all acetonitrile has evaporated (~2–3 μL final volume). Note to not completely overdry/dehydrate peptide sample due to issues with sample loss and resolubilization.

18. Add correspondingly a volume of Buffer A (0.1% TFA, 2% acetonitrile) in water to final volume 10 μL.
19. Withdraw in MS-specific vial for analysis (typically 3 μL loaded representing ~3.5–4 μg sample). To determine peptide concentration, a spectrophotometer analysis can be obtained based on 215 nm absorbance and comparison with known standard.
20. Short-term storage at 4 °C (within 2 weeks) or long-term −80 °C (up to 18 months).

MS/MS Analysis

1. Tryptic peptides were subjected to GeLC-MS/MS using a nanoflow UPLC instrument (Ultimate 3000 RSLCnano Thermo Fisher Scientific) coupled online to a high-resolution linear ion trap (Orbitrap Elite) mass spectrometer (Thermo Fisher Scientific) equipped with a nanoelectrospray ion source (Thermo Fisher Scientific).
2. Tryptic peptide samples were first loaded on a pre-column (Acclaim PepMap100 C18 5 μm 100 Å, Thermo Fisher Scientific) and separated using a VYDACMS C18-reversed phase column (25 cm length, 75 μm inner diameter, 3 μm 300 Å, Grace, Hesperia, CA) with a 120-min linear gradient from 0 to 100% (v/v) phase B (0.1% (v/v) FA in 80% (v/v) acetonitrile) at a flow rate of 250 nL/min [28].
3. MS data were acquired using a data-dependent Top 20 method dynamically choosing the most abundant precursor ions from the survey scan (300–2500 Th) using CID fragmentation. Survey scans were acquired at a resolution of 120,000 at *m*/*z* 400.
4. Unassigned precursor ion charge states as well as singly charged species were rejected, and peptide match was disabled.
5. The isolation window was set to 3 Th and fragmented with normalized collision energies of 25.
6. The maximum ion injection times for the survey scan and the MS/MS scans were 20 and 60 ms, respectively, and the ion target values were set to 3E6 and 1E6, respectively.
7. Selected sequenced ions were dynamically excluded for 90 s. Data were acquired using Xcalibur software v2.1 (Thermo Fisher Scientific).

Database Searching and Protein Identification

1. Raw data were processed using Proteome Discoverer (v1.4.0.288, Thermo Fisher Scientific). MS2 spectra were searched with Mascot (Matrix Science, London, UK; v 1.4.0.288), Sequest (Thermo Fisher Scientific, San Jose, CA, v 1.4.0.288), and X! Tandem (v 2010.12.01.1) against a database of 125,803 ORFs (UniProt Human, 2014_07).

2. Peptide lists were generated from a tryptic digestion with up to two missed cleavages, carbamidomethylation of cysteines as fixed modifications, and oxidation of methionines and protein N-terminal acetylation as variable modifications.

3. Precursor mass tolerance was 10 ppm, product ions were searched at 0.6 Da tolerances, min peptide length defined at 6, maximum peptide length 144, and max delta CN 0.05. Peptide spectral matches (PSM) were validated using Percolator based on *q*-values at a 1% false discovery rate (FDR) [51, 52].

4. With Proteome Discoverer, peptide identifications were grouped into proteins according to the law of parsimony and filtered to 1% FDR [53].

5. Scaffold (Proteome Software Inc., Portland, OR, v 4.3.4) was employed to validate MS/MS-based peptide and protein identifications from database searching. Initial peptide identifications were accepted if they could be established at greater than 95% probability as specified by the Peptide Prophet algorithm [54]. Protein probabilities were assigned by the Protein Prophet algorithm [53].

6. Protein identifications were accepted, if they reached greater than 99% probability and contained at least two identified unique peptides. These identification criteria typically established <1% false discovery rate based on a decoy database search strategy at the protein level. Proteins that contained similar peptides and could not be differentiated based on MS/MS analysis alone were grouped to satisfy the principles of parsimony. Contaminants and reverse identification were excluded from further data analysis. UniProt was used for protein annotation (cellular compartment, subcellular location, molecular function, transmembrane regions).

Label-Free Spectral Counting (*See Note 13*)

1. The relative abundance of a protein within a sample was estimated using semiquantitative normalized spectral count (Nsc) ratios. For each individual protein, significant peptide MS/MS spectra (i.e., ion score greater than identity score) were summated and normalized by the total number of significant MS/MS spectra identified in the sample (Eq. 1):

$$\text{Nsc} = (n + f) / (t - n + f) \quad (1)$$

where n is the number of significant peptide spectral counts for each protein in the sample, t is the total number of significant MS/MS spectral counts identified in the sample, and f is the correction factor set to 1.25. A correction factor is required to allow an Nsc value to be calculated when $n = 0$.

2. To compare relative protein abundance between samples, the ratio of normalized spectral counts (Rsc) was estimated (Eq. 2), as previously described based on studies by [55, 56]:

$$\mathrm{Rsc} = \left[(n_{\mathrm{B}} + f)(t_{\mathrm{A}} - n_{\mathrm{A}} + f)\right] / \left[(n_{\mathrm{A}} + f)(t_{\mathrm{B}} - n_{\mathrm{B}} + f)\right] \quad (2)$$

where n is the significant protein spectral count, t is the total number of significant MS/MS spectra in the sample, f is a correction factor set to 1.25, and A and B are the samples being compared.

4 Notes

1. Human colon carcinoma LIM1863 cells established at the Ludwig Institute in Melbourne from a resected invasive human colon carcinoma grow as free-floating suspension clusters that contain differentiated columnar cells. LIM1863 cells consistently yield a degree of morphological differentiation similar to that seen in the intestinal crypt and have many of the properties of the intestinal stem cell (stemness) [49]. This cell model has been extensively characterized by our laboratory [16, 18, 20, 28, 48].
2. For cell culture models, cell viability should be assessed to ensure that cells remain viable in serum-free culture conditions (refer notes in [48]).
3. Our group recently performed a proteomic analysis evaluating the ability of different techniques (namely, differential ultracentrifugation, OptiPrep density-gradient centrifugation, and immunocapture using EpCAM (CD326) antibodies coupled to magnetic beads) to enrich for exosome markers and proteins involved in exosome biogenesis, trafficking, and release. A detailed description [16] and protocol [48] describing these different methods for EV isolation are described.
4. Even though EVs are increasingly recognized as important biological and therapeutical entities, standardized methods for their analysis are still lacking. A detailed position paper on standardized sample collection, isolation, and analysis methods in EV research was recently reported [30]. Various recommendations on discovery research, characterization, and diagnostic research are discussed, defined, and refined by the research community [33–35]. In this protocol we have utilized a novel method to isolate EV subtypes from a cancer model and characterize these EVs using standard approaches. A salient finding was the first report of 350 proteins uniquely identified in sMVs, many of which have the potential to enable discrimination of this EV subtype from exosomes (notably, members of the

septin family, kinesin-like protein (KIF23), exportin-2/chromosome segregation like-1 protein (CSE1L), and Rac GTPase-activating protein 1 (RACGAP1)). These components have further been validated using conventional differential centrifugation, supporting their use as select sMV markers (Fig. 4b).

5. Sequential centrifugal ultrafiltration (SCUF). Low protein-binding PDVF ultrafilters were employed to reduce protein/EV interaction, clogging, and protein loss. Other size range filter membranes could be used to select other EV subtypes; however 98 % of protein content and EV detected from the five fractions in this protocol were from fractions 1 and 5 (sMVs and Exos, respectively). SCUF-Fn1, which contains large EVs (SCUF-sMVs), was centrifuged at 10,000 × *g* for 30 min, and it could reduce exosome fraction contamination.

6. EV characterization. There are numerous methods to characterize specific subtypes of EVs. In addition size, expression of specific proteins based on the biogenesis and cargo selection of EVs has been used [24, 57, 58]. In this protocol immunoblotting is used to verify expression of select protein markers: Alix (exosome formation, secretion), TSG101 (exosome formation, secretion), CD63 (exosome secretion), CD81 (exosome secretion), KIF23 (sMV marker), CSE1L (sMV marker), RACGAP1 (sMV marker), while EpCAM is commonly expressed in exosomes and sMVs. Although several other exosomal and sMV markers exist, these data provided herein refer specifically to the human LIM1863 cell model. CD63 and EpCAM should be used in nonreducing conditions for SDS-PAGE [28].

7. Buoyant density. In comparison with sucrose density gradients, OptiPrep velocity gradients have low viscosity, iso-osmotic gradients that provide rapid and efficient separation of EVs. OptiPrep velocity gradients have been shown to efficiently separate exosomes from HIV-1 particles [59]. These density gradients are often utilized by layering the CCM/crude EV mixtures on top of the sucrose gradient [60–62], a sucrose cushion [63, 64] or iodixanol [59] with centrifugation for a few hours to overnight. If centrifugation is too short and/or the tube is too long, though, contaminating aggregates may not reach the bottom of the tube and thus end up in the same fractions as EVs. Since different classes of EVs may have overlapping densities, however, even density-based procedures may achieve enrichment rather than true isolation. Exosomes have been shown to have buoyant density in the range of 1.09–1.12 g/mL (predominantly enriched at 1.10–1.11 g/mL) on continuous sucrose or iodixanol density gradients following ultracentrifugation [16, 17, 19]. While sMVs have been shown to float at buoyant densities in the range of 1.06–1.22 g/mL (unpublished observations), expression of KIF23 is predominantly detected in fraction 1.18–1.19 g/mL [28].

8. EV quantification. Standardized methods for EV analysis and quantification are still lacking [30]. Recent studies have shown the importance of technical knowledge of the instruments used for EV quantification to ensure that data interpretation is accurate [65]. Understanding that distinct EVs exist with different concentrations will ensure that users are not compromised by EV heterogeneity. Increased understanding of the possibilities and pitfalls of these technologies for characterization will benefit standardized and large-scale clinical application of EVs.
9. Dynamic light scattering (DLS). DLS allows the size distribution profile of small particles in suspension to be determined. For use in EV characterization, users should ensure a low EV protein concentration (typically 1–2 μg, with further dilutions often required). A concurrent issue with DLS is that often a small amount of larger particles or aggregates (high volume and displacement) can compromise results. Although multiple aspects and interpretation of the data could be obtained (i.e., intensity, volume, number distribution graphs), we recommend volume and number distribution for EV detection. Further DLS does not differentiate EVs from protein aggregates of similar size due to protein complexes, especially insoluble immune complexes, overlap in biophysical properties (size, light scattering, and sedimentation) [66–70].
10. Electron microscopy (EM). In this protocol, cryo-EM is provided although standard transmission EM could also be used to provide direct evidence for the presence of vesicular structures. Cyro-EM is recommended to measure Exos and sMVs based on high-resolution analysis of particle size and morphology. Cryo-EM utilizes EVs in frozen samples with the advantage of avoiding effects of dehydration and chemical fixation. Larger sMV membranes >1000 nm are difficult to detect, possibly associated with the stability of larger EV structures when frozen [28]. Further, cryo-EM allows tomographic data collection and the ability for spatial visualization of more complex structures.
11. Nanoparticle tracking analysis (NTA). NTA is an optical particle tracking method to measure the size distribution and concentration of EVs. It tracks the movement of laser-illuminated individual particles under Brownian motion and calculates their diameter using statistical models. An advantage of NTA is the unbiased high peak resolution for polydisperse samples. Like DLS, NTA does not differentiate EVs from protein aggregates of similar size [29, 66, 71]. Dilution is required for the NTA (10^6–10^9 particles per mL recommended), with further recommendations including 20–40 particles in the field of view and $2–10 \times 10^8$ particle concentration [69, 72–74]. A recent study has shown that the NTA camera level and detection threshold to be significant factors in the quantification of

liposomes, although changing these variable settings were less prominent for quantification of EVs [65].

12. In this protocol, given that conventional high-resolution mass spectrometers can utilize more complex, unfractionated (or limited fractionation) samples, we typically employ a short range gel (i.e., 9–10 mm into the gel well), with the gel piece divided into 2–3 equal portions. These portions are then individually reduced, alkylated, and tryptically digested [48].
13. In this chapter, label-free proteomic profiling was employed to deliver relative quantification of protein expression between distinct EVs. Spectral counting (SpC) counts the number of spectra identified for a given peptide in different biological samples and then integrates the results for all measured peptides of the protein(s) that are quantified [75]. SpC is a commonly used approach for measuring protein abundance in label-free proteomic analyses [48]. Further details on quantitative mass spectrometry relating to comparative analyses can be obtained from [76]. Other protocols can be used to focus on spectral counting quantification specifically [77], with relevance to extracellular vesicles [16–18, 78].

Acknowledgments

This work was supported, in part, by the La Trobe University Leadership RFA Grant (D.W.G, R.J.S), La Trobe Institute for Molecular Science Molecular Biology Fellowship (D.W.G), and La Trobe University Start-up Grant (D.W.G). RX is supported by La Trobe University Post Graduate Scholarship. We acknowledge the La Trobe University Comprehensive Proteomics Platform.

References

1. Xu R, Greening DW, Zhu HJ, Takahashi N, Simpson RJ. (2016) Extracellular vesicle isolation and characterization: toward clinical application. J Clin Invest. 126(4):1152–62
2. Valadi H, Ekstrom K, Bossios A et al (2007) Exosome-mediated transfer of mRNAs and microRNAs is a novel mechanism of genetic exchange between cells. Nat Cell Biol 9:654–659
3. van Niel G, Porto-Carreiro I, Simoes S et al (2006) Exosomes: a common pathway for a specialized function. J Biochem 140:13–21
4. Peinado H, Lavotshkin S, Lyden D (2011) The secreted factors responsible for pre-metastatic niche formation: old sayings and new thoughts. Semin Cancer Biol 21:139–146
5. Ratajczak J, Miekus K, Kucia M et al (2006) Embryonic stem cell-derived microvesicles reprogram hematopoietic progenitors: evidence for horizontal transfer of mRNA and protein delivery. Leukemia 20:847–856
6. Nazarenko I, Rana S, Baumann A et al (2010) Cell surface tetraspanin Tspan8 contributes to molecular pathways of exosome-induced endothelial cell activation. Cancer Res 70: 1668–1678
7. Webber J, Steadman R, Mason MD et al (2010) Cancer exosomes trigger fibroblast to myofibroblast differentiation. Cancer Res 70:9621–9630
8. Liu Y, Xiang X, Zhuang X et al (2010) Contribution of MyD88 to the tumor exosome-mediated induction of myeloid derived suppressor cells. Am J Pathol 176: 2490–2499

9. Xiang X, Poliakov A, Liu C et al (2009) Induction of myeloid-derived suppressor cells by tumor exosomes. Int J Cancer 124: 2621–2633
10. Al-Nedawi K, Meehan B, Micallef J et al (2008) Intercellular transfer of the oncogenic receptor EGFRvIII by microvesicles derived from tumour cells. Nat Cell Biol 10:619–624
11. Hao S, Ye Z, Li F et al (2006) Epigenetic transfer of metastatic activity by uptake of highly metastatic B16 melanoma cell-released exosomes. Exp Oncol 28:126–131
12. Skog J, Wurdinger T, van Rijn S et al (2008) Glioblastoma microvesicles transport RNA and proteins that promote tumour growth and provide diagnostic biomarkers. Nat Cell Biol 10:1470–1476
13. Greening DW, Gopal SK, Xu R et al (2015) Exosomes and their roles in immune regulation and cancer. Semin Cell Dev Biol 40:72–81
14. Greening DW, Gopal SK, Mathias RA et al (2015) Emerging roles of exosomes during epithelial-mesenchymal transition and cancer progression. Semin Cell Dev Biol 40:60–71
15. E-LA S, Mager I, Breakefield XO et al (2013) Extracellular vesicles: biology and emerging therapeutic opportunities. Nat Rev Drug Discov 12:347–357
16. Tauro BJ, Greening DW, Mathias RA et al (2012) Comparison of ultracentrifugation, density gradient separation, and immunoaffinity capture methods for isolating human colon cancer cell line LIM1863-derived exosomes. Methods 56:293–304
17. Ji H, Greening DW, Barnes TW et al (2013) Proteome profiling of exosomes derived from human primary and metastatic colorectal cancer cells reveal differential expression of key metastatic factors and signal transduction components. Proteomics 13:1672–1686
18. Tauro BJ, Greening DW, Mathias RA et al (2013) Two distinct population of exosomes released from LIM1863 colon carcinoma cells. Mol Cell Proteomics 12:587–598
19. Tauro BJ, Mathias RA, Greening DW et al (2013) Oncogenic H-ras reprograms Madin-Darby canine kidney (MDCK) cell-derived exosomal proteins following epithelial-mesenchymal transition. Mol Cell Proteomics 12:2148–2159
20. Ji H, Chen M, Greening DW et al (2014) Deep sequencing of RNA from three different extracellular vesicle (EV) subtypes released from the human LIM1863 colon cancer cell line uncovers distinct miRNA-enrichment signatures. PLoS One 9, e110314
21. Greening DW, Ji H, Chen M, Robinson BW, Dick IM, Creaney J, Simpson RJ. (2016) Secreted primary human malignant mesothelioma exosome signature reflects oncogenic cargo. Sci Rep. 6:32643
22. Greening DW, Nguyen HP, Elgass K, Simpson RJ, Salamonsen LA. (2016) Human Endometrial Exosomes Contain Hormone-Specific Cargo Modulating Trophoblast Adhesive Capacity: Insights into Endometrial-Embryo Interactions. Biol Reprod. 94(2):38
23. Cocucci E, Racchetti G, Meldolesi J (2009) Shedding microvesicles: artefacts no more. Trends Cell Biol 19:43–51
24. Clancy JW, Sedgwick A, Rosse C et al (2015) Regulated delivery of molecular cargo to invasive tumour-derived microvesicles. Nat Commun 6:6919
25. Pospichalova V, Svoboda J, Dave Z et al (2015) Simplified protocol for flow cytometry analysis of fluorescently labeled exosomes and microvesicles using dedicated flow cytometer. J Extracell Vesicles 4:25530
26. Di Vizio D, Kim J, Hager MH et al (2009) Oncosome formation in prostate cancer: association with a region of frequent chromosomal deletion in metastatic disease. Cancer Res 69:5601–5609
27. Antonyak MA, Wilson KF, Cerione RA (2012) R(h)oads to microvesicles. Small GTPases 3:219–224
28. Xu R, Greening DW, Rai A et al (2015) Highly-purified exosomes and shed microvesicles isolated from the human colon cancer cell line LIM1863 by sequential centrifugal ultrafiltration are biochemically and functionally distinct. Methods 87:11–25
29. Colombo M, Raposo G, Thery C (2014) Biogenesis, secretion, and intercellular interactions of exosomes and other extracellular vesicles. Annu Rev Cell Dev Biol 30:255–289
30. Witwer KW, Buzas EI, Bemis LT et al (2013) Standardization of sample collection, isolation and analysis methods in extracellular vesicle research. J Extracell Vesicles 2:24009894
31. Gyorgy B, Hung ME, Breakefield XO et al (2015) Therapeutic applications of extracellular vesicles: clinical promise and open questions. Annu Rev Pharmacol Toxicol 55:439–464
32. Nordin JZ, Lee Y, Vader P et al (2015) Ultrafiltration with size-exclusion liquid chromatography for high yield isolation of extracellular vesicles preserving intact biophysical and functional properties. Nanomedicine 11: 879–883
33. Gould SJ, Raposo G (2013) As we wait: coping with an imperfect nomenclature for

extracellular vesicles. J Extracell Vesicles 2:24009890

34. van der Pol E, Boing AN, Harrison P et al (2012) Classification, functions, and clinical relevance of extracellular vesicles. Pharmacol Rev 64:676–705
35. Mullier F, Bailly N, Chatelain C et al (2013) Pre-analytical issues in the measurement of circulating microparticles: current recommendations and pending questions. J Thromb Haemost 11:693–696
36. Thery C, Amigorena S, Raposo G et al (2006) Isolation and characterization of exosomes from cell culture supernatants and biological fluids. Curr Protoc Cell Biol 3:22
37. Grant R, Ansa-Addo E, Stratton D et al (2011) A filtration-based protocol to isolate human plasma membrane-derived vesicles and exosomes from blood plasma. J Immunol Methods 371:143–151
38. Merchant ML, Powell DW, Wilkey DW et al (2010) Microfiltration isolation of human urinary exosomes for characterization by MS. Proteomics Clin Appl 4:84–96
39. Lai RC, Arslan F, Lee MM et al (2010) Exosome secreted by MSC reduces myocardial ischemia/reperfusion injury. Stem Cell Res 4:214–222
40. Mathivanan S, Lim JW, Tauro BJ et al (2010) Proteomics analysis of A33 immunoaffinity-purified exosomes released from the human colon tumor cell line LIM1215 reveals a tissue-specific protein signature. Mol Cell Proteomics 9:197–208
41. Balaj L, Atai NA, Chen W et al (2015) Heparin affinity purification of extracellular vesicles. Sci Rep 5:10266
42. Rao S, Gefroh E, Kaltenbrunner O (2012) Recovery modeling of tangential flow systems. Biotechnol Bioeng 109:3084–3092
43. Petersen KE, Manangon E, Hood JL et al (2014) A review of exosome separation techniques and characterization of B16-F10 mouse melanoma exosomes with AF4-UV-MALS-DLS-TEM. Anal Bioanal Chem 406: 7855–7866
44. Lee C, Mitsialis SA, Aslam M et al (2012) Exosomes mediate the cytoprotective action of mesenchymal stromal cells on hypoxia-induced pulmonary hypertension. Circulation 126:2601–2611
45. Kordelas L, Rebmann V, Ludwig AK et al (2014) MSC-derived exosomes: a novel tool to treat therapy-refractory graft-versus-host disease. Leukemia 28:970–973
46. Shin H, Han C, Labuz JM et al (2015) High-yield isolation of extracellular vesicles using aqueous two-phase system. Sci Rep 5:13103
47. Chen C, Skog J, Hsu CH et al (2010) Microfluidic isolation and transcriptome analysis of serum microvesicles. Lab Chip 10:505–511
48. Greening DW, Xu R, Ji H et al (2015) A protocol for exosome isolation and characterization: evaluation of ultracentrifugation, density-gradient separation, and immunoaffinity capture methods. Methods Mol Biol 1295:179–209
49. Whitehead RH, Jones JK, Gabriel A et al (1987) A new colon carcinoma cell line LIM1863 that Grows as organoids with spontaneous differentiation into crypt like structures in vitro.pdf> Cancer Res 47:2683–2689
50. Schroder M, Schafer R, Friedl P (1997) Spectrophotometric determination of iodixanol in subcellular fractions of mammalian cells. Anal Biochem 244:174–176
51. Brosch M, Yu L, Hubbard T et al (2009) Accurate and sensitive peptide identification with Mascot Percolator. J Proteome Res 8:3176–3181
52. Greening DW, Kapp EA, Ji H et al (2013) Colon tumour secretopeptidome: insights into endogenous proteolytic cleavage events in the colon tumour microenvironment. Biochim Biophys Acta 1834:2396–2407
53. Nesvizhskii AI, Aebersold R (2005) Interpretation of shotgun proteomic data: the protein inference problem. Mol Cell Proteomics 4:1419–1440
54. Keller A, Nesvizhskii AI, Kolker E et al (2002) Empirical statistical model to estimate the accuracy of peptide identifications made by MS/MS and database search. Anal Chem 74:5383–5392
55. Beissbarth T, Hyde L, Smyth GK et al (2004) Statistical modeling of sequencing errors in SAGE libraries. Bioinformatics 20(Suppl 1):i31–i39
56. Old WM, Meyer-Arendt K, Aveline-Wolf L et al (2005) Comparison of label-free methods for quantifying human proteins by shotgun proteomics. Mol Cell Proteomics 4:1487–1502
57. Colombo M, Moita C, van Niel G et al (2013) Analysis of ESCRT functions in exosome biogenesis, composition and secretion highlights the heterogeneity of extracellular vesicles. J Cell Sci 126:5553–5565
58. Muralidharan-Chari V, Clancy J, Plou C et al (2009) ARF6-regulated shedding of tumor cell-derived plasma membrane microvesicles. Curr Biol 19:1875–1885
59. Cantin R, Diou J, Belanger D et al (2008) Discrimination between exosomes and HIV-1: purification of both vesicles from cell-free supernatants. J Immunol Methods 338:21–30

60. Bard MP, Hegmans JP, Hemmes A et al (2004) Proteomic analysis of exosomes isolated from human malignant pleural effusions. Am J Respir Cell Mol Biol 31:114–121
61. Keller S, Ridinger J, Rupp AK et al (2011) Body fluid derived exosomes as a novel template for clinical diagnostics. J Transl Med 9:86
62. Poliakov A, Spilman M, Dokland T et al (2009) Structural heterogeneity and protein composition of exosome-like vesicles (prostasomes) in human semen. Prostate 69: 159–167
63. van Balkom BW, Pisitkun T, Verhaar MC et al (2011) Exosomes and the kidney: prospects for diagnosis and therapy of renal diseases. Kidney Int 80:1138–1145
64. Lamparski HG, Metha-Damani A, Yao JY et al (2002) Production and characterization of clinical grade exosomes derived from dendritic cells. J Immunol Methods 270:211–226
65. Maas SL, de Vrij J, van der Vlist EJ et al (2015) Possibilities and limitations of current technologies for quantification of biological extracellular vesicles and synthetic mimics. J Control Release 200:87–96
66. Filipe V, Hawe A, Jiskoot W (2010) Critical evaluation of nanoparticle tracking analysis (NTA) by nanosight for the measurement of nanoparticles and protein aggregates. Pharm Res 27:796–810
67. Genneback N, Hellman U, Malm L et al (2013) Growth factor stimulation of cardiomyocytes induces changes in the transcriptional contents of secreted exosomes. J Extracell Vesicles 2:PMCID: PMC3760655
68. Kesimer M, Gupta R (2015) Physical characterization and profiling of airway epithelial derived exosomes using light scattering. Methods 87:59–63
69. Sitar S, Kejzar A, Pahovnik D et al (2015) Size characterization and quantification of exosomes by asymmetrical-flow field-flow fractionation. Anal Chem 87:9225–9233
70. Gyorgy B, Modos K, Pallinger E et al (2011) Detection and isolation of cell-derived microparticles are compromised by protein complexes resulting from shared biophysical parameters. Blood 117:e39–e48
71. Gardiner C, Ferreira YJ, Dragovic RA et al (2013) Extracellular vesicle sizing and enumeration by nanoparticle tracking analysis. J Extracell Vesicles 2:PMCID: PMC3760643
72. Bisaro B, Mandili G, Poli A et al (2015) Proteomic analysis of extracellular vesicles from medullospheres reveals a role for iron in the cancer progression of medulloblastoma. Mol Cell Ther 3:8
73. Jeppesen DK, Hvam ML, Primdahl-Bengtson B et al (2014) Comparative analysis of discrete exosome fractions obtained by differential centrifugation. J Extracell Vesicles 3:25011
74. Dragovic RA, Gardiner C, Brooks AS et al (2011) Sizing and phenotyping of cellular vesicles using nanoparticle tracking analysis. Nanomedicine 7:780–788
75. Asara JM, Christofk HR, Freimark LM et al (2008) A label-free quantification method by MS/MS TIC compared to SILAC and spectral counting in a proteomics screen. Proteomics 8:994–999
76. Griffin NM, Yu J, Long F et al (2010) Label-free, normalized quantification of complex mass spectrometry data for proteomic analysis. Nat Biotechnol 28:83–89
77. Arike L, Peil L (2014) Spectral counting label-free proteomics. Methods Mol Biol 1156: 213–222
78. Amorim M, Fernandes G, Oliveira P et al (2014) The overexpression of a single oncogene (ERBB2/HER2) alters the proteomic landscape of extracellular vesicles. Proteomics 14:1472–1479

Chapter 8

Multiplexed Phenotyping of Small Extracellular Vesicles Using Protein Microarray (EV Array)

Rikke Bæk and Malene Møller Jørgensen

Abstract

The Extracellular Vesicle (EV) Array is based on the technology of protein microarray and provides the opportunity to detect and phenotype small EVs from unpurified starting material in a high-throughput manner (Jørgensen et al., J Extracell vesicles 2:1–9, 2013). The technology was established to perform multiplexed phenotyping of EVs in an open platform. This protocol outlines the microarray printing procedure followed by the steps of capture and detection of small extracellular vesicles from plasma/serum or cell culture supernatants. The principles of data treatment and analysis are thoroughly described as well.

Key words Protein microarray, Extracellular vesicles, Exosomes, Capture antibodies, Surface antigens, Multiplexed phenotyping

Abbreviations

dAF488	Dextran coupled with Alexa Fluor 488
sEV	small Extracellular vesicle
IgG	Immunoglobin gamma
ON	Overnight
PBS	Phosphate-buffered saline
RT	Room temperature

1 Introduction

Extracellular vesicles (EVs) are released from healthy as well as diseased cells to facilitate cellular communication. They have a wide variety of names including exosomes, microvesicles, and microparticles. Several methods exist to characterize the protein composition of EVs either related to a surface marker phenotype or to the proteins present in the EV cargo as reviewed by Revenfeld et al. [1]. While the EV phenotype is particularly important in the determination of cellular and subcellular origin, it can in

Andrew F. Hill (ed.), *Exosomes and Microvesicles: Methods and Protocols*, Methods in Molecular Biology, vol. 1545, DOI 10.1007/978-1-4939-6728-5_8,

combination with a protein cargo analysis also provide clues about the functionality of the EVs.

The EV Array is based on the technology of protein microarray and allows the detection and phenotyping of sEVs from unpurified starting material in a high-throughput manner [2]. The technology was developed to perform multiplexed phenotyping of sEVs in an open platform. Protein microarrays are well accepted as powerful tools in the search for antigens or antibodies in various sample types [3, 4] and constitutes basis for a high-throughput method. This provides the opportunity to track interactions and activities of proteins in a large scale. A great advantage of protein microarray is that a large number of proteins can be tracked in parallel. Using the EV Array, it is at present possible to phenotype for at least 60 markers simultaneously [5]. In addition, the technique offers a rapid and highly sensitive method consuming only small quantities of samples and reagents. An outline of the laboratory procedures, requirements, and time frames for the EV Array is illustrated in Fig. 1a.

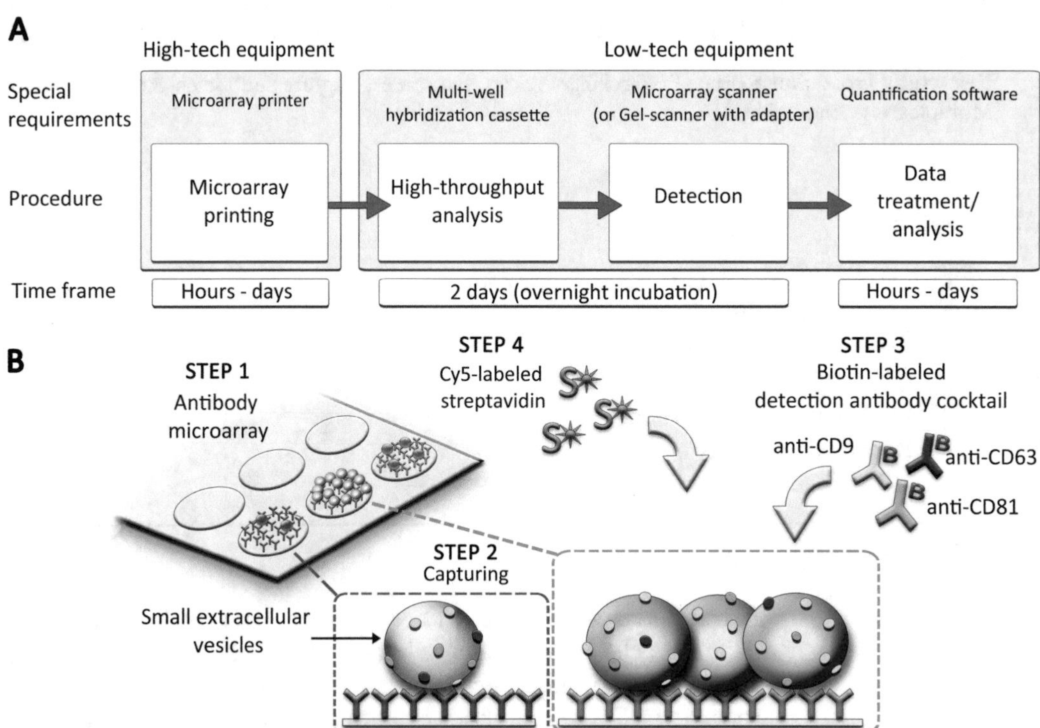

Fig. 1 (**a**) Schematic view of the laboratory procedures, requirements, and time frames for the EV Array. (**b**) Graphical sketch of the EV Array procedure. *STEP 1*: Epoxy-silane-coated microarray slides are the basis for the EV Array. *STEP 2*: EV containing samples (plasma/serum or cell culture supernatants) are applied, and the sEVs are captured onto the slides depending on their specific surface antigens. *STEP 3*: A cocktail of biotin-labeled detection antibodies is applied. Here illustrated with a cocktail of antibodies against the exosomal markers CD9, CD63, and CD81. *STEP 4*: Fluorescently labeled streptavidin is added prior to detection of the captured sEVs. *The time needed for printing and data analysis depends on the number of analytes and samples to be analyzed

Micro-sized spots of capturing antibodies against known EV surface antigens are printed in a customized setup using microarray printing technology (Fig. 1b STEP 1). The EV Array is printed on standard epoxysilane-coated microarray slides in patterns fitting into each well of a multi-well cassette. The use of a 96-well multi-well cassette allows the EV Array to be performed as a high-throughput procedure using a minimum of sample. A maximum of 100 μL of vesicle-containing sample can be applied to each well (Fig. 1b STEP 2). The captured vesicles are detected using a cocktail of biotinylated antibodies against the tetraspanins CD9, CD63, and CD81 (Fig. 1b STEP 3). These markers are regarded as exosomal markers in general [6–8], which is the type of EV that the EV Array is designed to analyse, though other sEVs of the same size are captured as well. This particular antibody combination is applied to ensure that all captured exosomes are detected. If wanted, the focus can easily be changed from exosomes to another type of sEV by replacing the detection antibodies, though subsequent optimization may be needed. Afterward, fluorescently labeled streptavidin is used to quantify the amount of exosomes captured on each spot (Fig. 1b STEP 4). The readout is a fluorescence signal for each individual microarray spot. The method is semiquantitative and determines the phenotype of sEVs for up to 60 protein markers simultaneously.

The EV Array is optimized as a high-throughput analysis where 20 samples can be analyzed simultaneously on each microarray slide. The technique features an overnight (ON) incubation to capture the sEVs and consequently the method can be completed in 2 days. As outlined, the development of the EV Array technology leaves the possibilities open to simply change the detection antibodies in order to detect other populations or subpopulations of sEVs as, e.g., tissue factor-bearing vesicles (Fig. 1b STEP 3). Hereby an open platform is generated, which is easy to transform and optimize further in a research and clinical perspective.

As clarified in Fig. 1a, specific laboratory requirements are necessary to perform an EV Array. The microarray epoxy-coated slides, which are custom-printed with the antibodies against the targeted EV marker proteins, are the basis of the Array. For laboratories without access to microarray printing technologies, several commercial companies offer custom-made microarray slides. To use the EV Array in a high-throughput manner, a simple multi-well cassette to allow multiple Arrays on each slide is required, and to detect the fluorescent signals, either a microarray scanner or laboratory gel scanner is needed.

Using a method such as the EV Array to phenotype sEVs has shown a great potential in diagnostics as highlighted by Jakobsen et al. [9] and Sandfeld-Paulsen et al. [10] where the EV Array was used to phenotype plasma sEVs from NSCLC patients and other lung-diseased controls. By establishing a multi-marker model, it was possible to separate the cancer patients with up to 75.3% accuracy.

2 Materials

Prepare all solutions using purified deionized water (sensitivity of 18 MΩ cm at 25 °C) and analytical grade reagents. Prepare and store all reagents at room temperature (RT) (unless otherwise indicated). When handling the microarray slides, powder-free gloves should be used and touching should be limited to the edges of the slide.

2.1 Printing of Microarray

1. SpotBot Extreme Protein Microarray Spotter (Arrayit Corporation, CA, USA).
2. 946MP4 Pin (Arrayit Corporation, CA, USA).
3. Nexterion® Slide E, epoxysilane-coated glass slides with barcode (SCHOTT Nexterion, Germany).
4. Phosphate-buffered saline (PBS): 137 mM NaCl, 10 mM Na_2HPO_4, 2.7 mM KCl, 1.8 mM KH_2PO_4, and adjust to pH 7.4 with HCl. Sterile filter with 0.22 μm filter.
5. Glycerol, minimum 99% (*see* **Note 1**).
6. Dextran coupled with Alexa Fluor 488 (dAF488), 10.000 MW, anionic, fixable (Life Technologies Corp, USA) (*see* **Note 2**).
7. Capture antibodies of interest: Diluted to 50–300 μg/mL in PBS containing a final glycerol content of 5% and 50 μg/mL dAF488 (store at –40 °C).
8. Positive control: 50–100 μg/mL biotinylated IgG in PBS containing a final glycerol content of 5% and 50 μg/mL dAF488 (store at −40 °C).
9. Negative control: PBS containing 5% glycerol and 50 μg/mL dAF488 (Store at –40 °C).
10. Microarray 384-well microplate (Arrayit Corporation, CA, USA).
11. InnoScan 710 AL microarray scanner (Innopsys, France).
12. Mapix (microarray analysis software, Innopsys, France).

2.2 EV Array Analysis

1. Peristaltic pump.
2. High-Throughput Wash Station (Arrayit Corporation, CA, USA).
3. Magnetic stirrer.
4. Blocking buffer: 50 mM ethanolamine, 100 mM Tris, 0.1% (v/v) SDS, pH 9.0. Add 3.05 mL ethanolamine solution, 12.1 g Tris, and 5 mL of 20% SDS solution to approximately 900 mL of water. Mix and adjust pH with NaOH. Add water up to a total of 1 L solution. Sterile filter with 0.22 μm and store at 4 °C.

5. Wash buffer: PBS (*see* Subheading 2.1) containing 0.05% Tween®20 (500 μL per 1 L PBS). Sterile filter with 0.22 μm and store at 4 °C.
6. Multi-well microarray hybridization cassette 4×24 or 1×24 (Arrayit Corporation, CA, USA).
7. Adhesive tape for ELISA plates or similar.
8. Orbital shaker.
9. Rectangular plate, four chambers, transparent, non-treated (Nunc, Thermo Fisher Scientific, MA, USA).
10. Detection antibodies (anti-human CD9/biotin, CD63/biotin, and CD81/biotin, Ancell Corporation, MN, USA; Catalog # 156-030, 215-030, and 302-030, respectively). Antibodies are diluted freshly at 1:1500 with wash buffer (3 mL per slide).
11. Streptavidin Cy®5 conjugate (Life Technologies Corp, USA): The streptavidin Cy®5 conjugate is diluted freshly at 1:1500 in 3 mL wash buffer per slide.
12. Microarray High-Speed Centrifuge (Arrayit Corporation, CA, USA).
13. InnoScan 710 AL microarray scanner (Innopsys, France).
14. Mapix (microarray analysis software, Innopsys, France).

3 Methods

3.1 Printing of Microarray

1. Place the prepared printing solutions in a microarray microplate; 10 μL antibody or control solutions are sufficient.
2. Give the plate a short spin (1 min) to remove air bubbles (*see* **Note 3**).
3. Place the microplate, epoxy-coated slides and one contact-printing pin in the SpotBot Extreme Microarray Printer and manufacture the EV Arrays using the 21-well Multiple Microarray Format function (*see* Fig. 2a and **Note 4**).
4. Keep the temperature at 15–18 °C and the humidity at 55–65% during the printing procedure by using the appertaining cooling bath and humidity controller, respectively (*see* **Note 5**).
5. Use an uncoated glass slide for disposal of 20 preprints after which each printing solution is printed as triplicate in each subarray (*see* Fig. 2b and **Note 6**).
6. Before applying the next antibody solution, perform a washing procedure on the pin, as recommended by the manufacturer.
7. Quality control of the printing is manually performed using the InnoScan microarray scanner and the Mapix software with the settings of 532 nm at 10 V, PTM at 100%, and a 20 μm resolution (*see* Fig. 2c and **Note 7**).

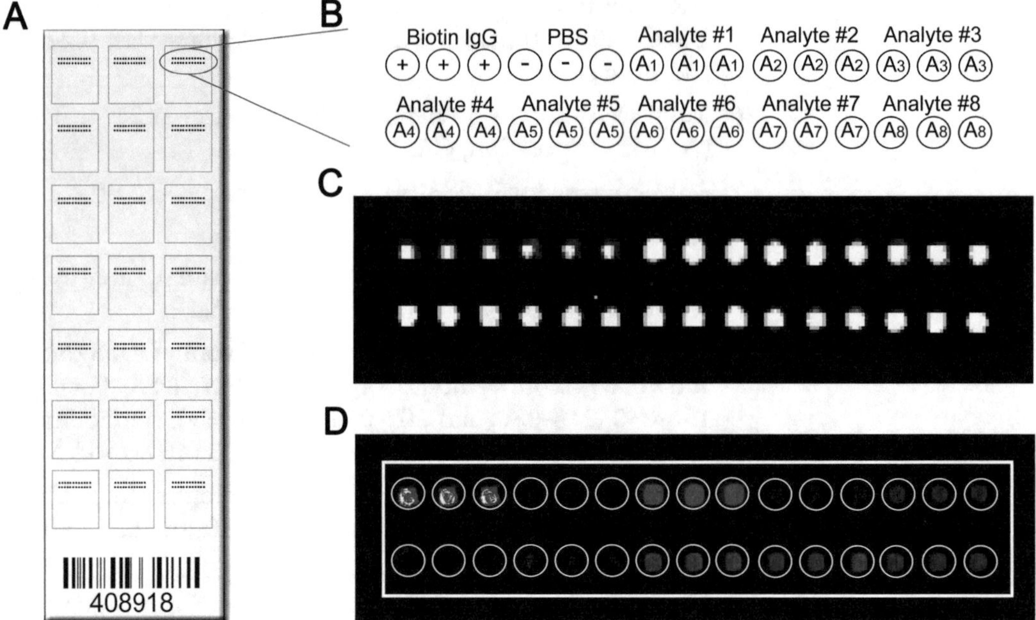

Fig. 2 (**a**) Bar-coded microarray glass slide (7.5 × 2.5 cm) printed in a 21-well setup as sketched out. (**b**) Outline of the printed analytes and controls in triplicates. Here sketched out with eight different analytes. (**c**) Quality control of the printed controls and analytes. Prior to analysis, the quality is validated by scanning at a resolution of 20 μm at 532 nm, a setting of 10 V, and a PMT at 100 %. (**d**) Scanning after capture and detection of sEVs. The scanning is performed at 635 nm, a setting of 10 V, and a PMT of 60 %. During data analysis a created *.GAL file is used to locate the printed spots as illustrated with the *white frame* and *spot circles*

8. Use the SpotBot software to generate a *.GAL file for the data analysis (*see* **Note 8**).
9. The EV Array slides are stored in closed containers at RT ON or longer until performance of the analysis (*see* **Note 1**).

3.2 EV Array Analysis

1. Place the EV Array slides into the gasket in the High-Throughput Wash Station placed on a magnetic stirrer set to 120 rpm.
2. Add 450 mL blocking buffer using the peristaltic pump at a speed of 5 mL/min (*see* **Note 9**).
3. After addition of blocking buffer, the slides are left to block for additional 30 min.
4. Exchange the blocking buffer with 450 mL of wash buffer and set the magnetic stirrer to 200 rpm, which is the setting for the remainder of the analysis.
5. Let the slides wash for 15 min.
6. Place the slides in the hybridization cassette (up to four slides per cassette) (*see* **Note 10**).
7. Load the samples of interest in the individual wells. Samples should be diluted in wash buffer (*see* **Notes 11** and **12**).

8. One well per slide should contain a blank/negative control (wash buffer) (*see* **Note 13**).
9. Seal the hybridization cassette with self-adhesive tape or similar to avoid evaporation.
10. Place the hybridization cassette on an orbital shaker set to 450 rpm for 2 h at RT and afterward incubate the cassette at 4 °C ON (no agitation is needed).
11. Empty the casette and carefully remove the slides from the hybridization cassette and place them in the High-Throughput Wash Station filled with wash buffer for 15 min.
12. Place the slides in the rectangular four chambers plate (one slide in each chamber) together with 3 mL of the detection antibody cocktail per chamber.
13. Place the chamber on an orbital shaker and incubate for 2 h at RT.
14. Wash the slides in the High-Throughput Wash Station for 15 min.
15. Place the slides in another rectangular four chambers plate with 3 mL of the diluted streptavidin Cy®5 conjugate per slide.
16. Incubate on an orbital shaker for 30 min under cover at RT (*see* **Note 14**).
17. Wash the slides in wash buffer for 15 min under cover, exchange the wash buffer with water and wash for additionally 15 min.
18. Dry the slides using the Microarray High-Speed Centrifuge for 20–30 s.
19. Load the slides into the microarray scanner (*see* **Note 15**).
20. Scan the slides at a resolution of 5 μm at 635 nm using 10 V and a PMT at 60%.
21. Open the *.tif file together with the respective *.GAL file in the Mapix software.
22. Locate all the printed spots either manually or automatically (*see* **Note 16**) using a fixed spot size (e.g., Ø135 μm).
23. Calculate the total spot intensity for each spot and use the save function to generate a *.txt file (*see* Fig. 3).
24. Import the *.txt file contents into Microsoft Excel or another mathematical program, and use the values of total intensity to calculate the relative intensity (RI) for each analyte and each sample (block) (e.g., *see* **Note 17** of calculation).

$$RI_{\text{Analyte}_y;\text{Block}_x} = \text{Log2}\left(\frac{\dfrac{\sum \text{Analyte}_y \text{Block}_x}{N} - \dfrac{\sum \text{Analyte}_y \text{Block}_{\text{Blank}}}{N}}{\dfrac{\sum \text{neg control, Block}_x}{N}} \right)$$

where N indicates the number of spot replicates (triplicates).

Block	Column	Row	ID	Name	F635 Total Intensity
1	1	1	Control	pos	13811532
1	2	1	Control	pos	14673881
1	3	1	Control	pos	15579693
1	4	1	Control	neg	49133
1	5	1	Control	neg	45979
1	6	1	Control	neg	52196
1	7	1	Analyte	#1	377353
1	8	1	Analyte	#1	421072
1	9	1	Analyte	#1	426949
1	10	1	Analyte	#2	91207
1	11	1	Analyte	#2	92161
1	12	1	Analyte	#2	98970
1	13	1	Analyte	#3	106520
1	14	1	Analyte	#3	109357
1	15	1	Analyte	#3	116220
⋮	⋮	⋮	⋮	⋮	⋮
21	7	1	Analyte	#1	5685
21	8	1	Analyte	#1	5426
21	9	1	Analyte	#1	5549
⋮	⋮	⋮	⋮	⋮	⋮

Fig. 3 Example of the content of a *.txt file generated after analysis of the EV Array. *Column 1* designates the block number/sample well in the 21-well format. *Columns 2* and *3* indicate the spot placement inside the block. *Columns 4* and *5* give the ID and name of the printed analyte (as indicated in the *.GAL file). The *last column* gives the total intensity of the spot scanned at 635 nm

4 Notes

1. The addition of glycerol ensures that the printed antibodies will stay in solution on the EV Array slides for up to one month at RT. Dried out spots will alter the shape of the processed spots and the signal cannot be accounted as reliable.
2. Addition of dAF488 provides the possibility to perform a quality control of the printed spots using the InnoScan microarray scanner. The dextran molecules do not adhere to the epoxy-coated slides; hence, it will not disturb the following analysis.
3. The centrifuge used is an Eppendorf 5810 centrifuge with rotor A-4-62 and a microplate adapter.
4. The 21-well Multiple Microarray Format function is a software setup that allows for subarray prints, which fit into the 4 × 24 (or 1 × 24) multi-well hybridization cassette used in the following EV Array analysis. The use of the 21-well setup instead of the 24-well setup is chosen to avoid printing on the barcode, which may cause the very delicate pins to block. Slides without barcode can be used and will allow printing in a 24-well setup.
5. If the humidity exceeds 75%, there is a major risk that the printing compartment will steam up, which will cause the spots

to blur together. The limits of the steam formation may vary depending on the humidity and temperature parameters in the surroundings.

6. The chosen pin type will deliver approx. 1 nL solution in each spot. The diameter of the spots is approx. 135 μm though it may vary depending on the constituents of the used antibody.
7. Pins used for microarray contact printing are very delicate due to the tiny sample channels, which in the given pin type will take up approx. 250 nL printing solution. Therefore, even though the printing compartment is a closed box and even if great care is taken when the printing is prepared, small dust particles will find their way into the compartment when placed in a non-clean room lab. This will occationally disturb the spotting procedure and therefore needs to be outlined. Scanning at 532 nm makes it easy to point out if spots are missing and, consequently, the print will be discarded in the following analysis.
8. The GenePix Array List (GAL) file describes the size and position of blocks (one per sample well) and the layout of analytes (spots) printed in them, together with the names and identifiers of the printed spots. The *.GAL files can also be generated manually using, e.g., Microsoft Excel.
9. Addition of blocking buffer is a critical step; if the blocking solution is added too fast, it will result in deformed spots (rocket shaped), which may interfere with each other and make the analysis unusable. In the following steps of the analysis, it is no longer necessary to add buffer in this gentle manner.
10. Make sure the hybridization cassette is closed thoroughly or leak may occur. Avoid the slides to dry out on the printed side at any time during the analysis. This will cause a high background signal/fluorescence to appear.
11. Plasma/serum samples: In most cases 10 μL of sample in 90 μL wash buffer is the best compromise of signal versus background. In some cases (low expressed analytes), it is relevant to add more sample, e.g., 20 μL sample and 80 μL wash buffer, but one should note that increasing the amount of sample most likely will increase the background signal, which is described in [5]. The use of different blood collection tubes and the effect of various preanalytical treatments is outlined in [11]. Furthermore, other kinds of bodyfluids have been tested as well and is described in [12].
12. Cell culture supernatant: If the supernatant of a given cell line is to be analyzed for the first time, it is recommended to concentrate it 20–100 times using a 100 K filter (e.g., Amicon Ultra-15 centrifugal filter unit with Ultracel-100 membrane, Merck Millipore). When loaded onto the EV Array, it should be diluted at least 1:1 in wash buffer depending on its concentration. A negative control/culture media control should be included.

When the given cell culture supernatant is characterized 100 µL of non-concentrated, undiluted sample is sufficient. Usually only a small subset of markers is positive when analyzing cell culture supernatant.

13. To streamline the data analysis, it is recommended to use the same well as the negative control (wash buffer) each time. As a standard well number 21 has been used.
14. To avoid the Cy5® fluorescence signal to decay, the slides should be protected from light during the incubation and washing steps.
15. A normal gel scanner (e.g., Typhoon, GE Healthcare) or similar fluorescence scanner can be used instead of the specialized microarray scanner.
16. Using automatically detection of blocks and spots a manually validations is necessary.
17. For this example (*see* Fig. 3), the negative control (wash buffer) is placed in well 21.

$$RI_{\text{Analyte}_{\#1};\text{Block}_1} = \text{Log}2\left(\frac{\frac{\sum \text{Analyte}_{\#1}\text{Block}_1}{N} - \frac{\sum \text{Analyte}_{\#1}\text{Block}_{21}}{N}}{\frac{\sum \text{neg control, Block}_1}{N}}\right)$$

$$RI_{\text{Analyte}_{\#1};\text{Block}_1} = \text{Log}2\left(\frac{\frac{(377.353+421.072+426.949)}{3} - \frac{(5.685+5.426+5.549)}{3}}{\frac{(49.133+45.979+52.196)}{3}}\right)$$

$$RI_{\text{Analyte}_{\#1};\text{Block}_1} = \text{Log}2\left(\frac{408.458-5.553}{49.103}\right) = 3.04$$

Acknowledgments

The authors would like to thank M.D. Kim Varming and Dr. Evo K.L. Søndergaard for helping establish this protocol.

References

1. Revenfeld ALS, Bæk R, Nielsen MH et al (2014) Diagnostic and prognostic potential of extracellular vesicles in peripheral blood. Clin Ther 36:830–846. doi:10.1016/j.clinthera.2014.05.008
2. Jørgensen M, Bæk R, Pedersen S et al (2013) Extracellular vesicle (EV) Array: microarray capturing of exosomes and other extracellular vesicles for multiplexed phenotyping. J Extracell Vesicles 2:1–9. doi:10.3402/jev.v2i0.20920

3. Melton L (2004) Proteomics in multiplex. Nature 429:101–107
4. Hall DA, Ptacek J, Snyder M (2007) Protein microarray technology. Mech Ageing Dev 128:161–167
5. Jørgensen M, Bæk R, Varming K (2015) Potentials and capabilities of the extracellular vesicle (EV) Array. J Extracell Vesicles 4:1–8
6. Glickman JN, Morton PA, Slot JW et al (1996) The biogenesis of the MHC class II compartment in human I-cell disease B lymphoblasts. J Cell Biol 132:769–785
7. Peters PJ, Neefjes J, Oorschot V et al (1991) Segregation of MHC class II molecules from MHC class I molecules in the Golgi complex for transport to lysosomal compartments. Nature 349:669–676
8. Raposo G, Nijman HW, Stoorvogel W et al (1996) B Lymphocytes secrete antigen-presenting vesicles. J Exp Med 183:1161–1172
9. Jakobsen KR, Paulsen BS, Bæk R et al (2015) Exosomal proteins as potential diagnostic markers in advanced non-small cell lung carcinoma. J Extracell Vesicles 4:26659
10. Sandfeld-Paulsen B, Jakobsen KR, Bæk R, Folkersen BH et al (2016) Exosomal Proteins as Diagnostic Biomarkers in Lung Cancer. J Thorac Oncol 11(10):1701–1710
11. Bæk R, Søndergaard EKL, Varming K, Jørgensen MM (2016) The impact of various preanalytical treatments on the phenotype of small extracellular vesicles in blood analyzed by protein microarray. J Imm Meth 438:11–20
12. Jørgensen MM, Bæk R, Tveito S et al (2016) Phenotyping of Small Extracellular Vesicles from Clinically Important Body Fluids using Protein Microarray. Book Chapter in "Exosomes: Biogenesis, Therapeutic Applications and Emerging Research", edited by Arlando Meyers, Nova Publishers

Chapter 9

Purification and Analysis of Exosomes Released by Mature Cortical Neurons Following Synaptic Activation

Karine Laulagnier, Charlotte Javalet, Fiona J. Hemming, and Rémy Sadoul

Abstract

Exosomes are vesicles released by most cells into their environment upon fusion of multivesicular endosomes with the plasma membrane. Exosomes are vesicles of 60–100 nm in diameter, floating in sucrose at a density of ~1.15 g/mL and carrying a number of marker proteins such as Alix, Tsg101, and Flotillin-1. We use dissociated cortical neurons cultured for around two weeks as exosome-releasing cells. In these conditions, neurons make mature synapses and form networks that can be activated by physiological stimuli. Here, we describe methods to culture differentiated cortical neurons, induce exosome release by increasing glutamatergic synapse activity, and purify exosomes by differential centrifugations followed by density separation using sucrose gradients. These protocols allow purification of neuronal exosomes released within minutes of activation of glutamatergic synapses.

Key words Exosomes, Mature neurons, Primary neuronal culture, Regulated secretion, Glutamatergic synapses, Bicuculline

1 Introduction

In the following protocols, "exosomes" refer to nanovesicles released in the extracellular medium upon fusion of multivesicular endosomes with the plasma membrane. Extracellular vesicles are called exosomes when (1) their diameter ranges from 60 to100 nm (observed by electron microscopy or nanoparticle tracking analysis), (2) they float at a density ranging from 1.1 to 1.19 g/mL on a linear gradient of sucrose, and (3) they carry typical protein markers such as Alix, Flotillin-1, Tsg101, or CD63 [1, 2]. An updated profile of proteins, lipids, and nucleic acids found in exosomes is available at www.microvesicles.org [3]. During the last 10 years, vesicles having the above-described characteristics of exosomes have been detected in different biological fluids including milk, saliva [4], and cerebrospinal fluid [5]. Exosomes are known to be released constitutively by various cell lines and can be purified from cell supernatants after several days of culture [6, 7].

Andrew F. Hill (ed.), *Exosomes and Microvesicles: Methods and Protocols*, Methods in Molecular Biology, vol. 1545,
DOI 10.1007/978-1-4939-6728-5_9, © Springer Science+Business Media LLC 2017

Concerning the central nervous system (CNS), several studies have shown that cultured astrocytes [8], microglia [9], and neurons [10] also secrete exosomes constitutively. These reports have led to the hypothesis that exosomes are involved in intercellular communications in the normal or pathological CNS [11, 12]. This hypothesis has been reinforced by the demonstration that exosomes released by myelinating cells (oligodendrocytes in the CNS and Schwann cells in the peripheral nervous system) can be endocytosed by neurons [13, 14]. A few studies have described exosome release within minutes in a regulated way following a rise of intracellular calcium due to calcium ionophores [15, 16] or activation of specific membrane receptors [17]. We have used neurons dissociated from rat cortices and cultured for 2 weeks in order to investigate how the release of exosomes might be regulated by neuronal activity. These cell cultures contain both excitatory glutamatergic and inhibitory GABAergic neurons, which make functional networks by the second week in culture. Thus, incubation with GABA receptor antagonists, such as picrotoxin or bicuculline, alleviates inhibitory activities within the networks and increases the excitatory activity of glutamatergic synapses [18, 19]. We have already reported that picrotoxin or bicuculline rapidly (10–15 min) and massively augmented the calcium-dependent secretion of exosomes in a way dependent on AMPA and NMDA receptors [20]. The first part of this chapter describes our protocol for the preparation of viable primary cultures of neurons up to their mature state. The second part explains how to stimulate exosome release using bicuculline. Finally, we give details for purifying exosomes from supernatants of stimulated neurons: this is based on low-speed centrifugations required to remove cell debris, followed by ultracentrifugation to pellet the nanovesicles. The vesicles are then purified on a linear 8–60% sucrose gradient where they move to a density of 1.1–1.15 g/mL. This protocol allows the characterization of a homogeneous exosome population, whose release is tightly regulated by glutamatergic activity (Fig 1).

2 Materials

2.1 Primary Cultures of Cortical Neurons

1. Animals: Use one pregnant rat (OFA strain) at embryonic day 18 (usually ~12–13 embryos) for one culture.
2. Borate buffer 0.1 M, pH 8.4: Dissolve 3.1 g boric acid (H_3BO_3) and 4.75 g sodium tetraborate (borax) in 1 L of distilled water. Adjust to pH 8.4. Filter the solution (0.22 μm) and store at 4 °C.
3. Coating solution: Dissolve poly-D-lysine (PDL, Sigma, molecular weight 30,000–70,000) in borate buffer in order to obtain a 50 μg/mL solution.

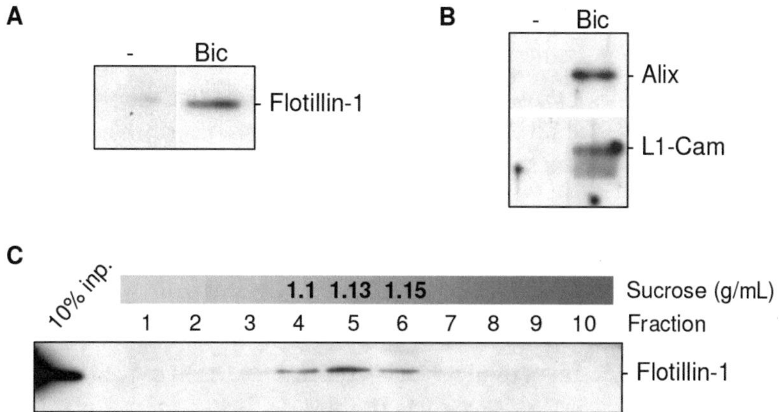

Fig. 1 Release of exosomes by mature neurons under glutamatergic stimulation. 52×10^6 cortical neurons were grown with (**a**) or without (**b**) AraC to DIV15. As described in Subheading 3.2, neurons were incubated in Neurobasal® medium alone ("–") or with 40 μM bicuculline ("Bic") for 15 min at 37 °C. Exosomes were pelleted from supernatants by ultracentrifugation. Pellets were resuspended in sample buffer, run on SDS-PAGE, and blotted against exosomal markers Flotillin-1 (50 kDa) and Alix (92 kDa) or the neuronal marker L1-Cam (200 kDa). Synaptic activation by bicuculline increases the amount of exosomes released by neurons as revealed by the increase in the detection of exosomal markers. (**c**) 52×10^6 neurons were treated with bicuculline-containing medium. Exosomes were pelleted from supernatants by ultracentrifugation and resuspended in 8 % sucrose solution. 10 % of the sample was run and blotted directly ("10 % inp."). The remaining 90 % were loaded onto a 8–60 % sucrose linear gradient and ultracentrifuged. Ten fractions of the gradient were collected and washed by ultracentrifugation, and pellets were resuspended in sample buffer for Western blot analysis. The exosomal marker Flotillin-1 appears in three fractions ranging from 1.1 to 1.15 g/mL of sucrose as expected for exosomes

4. Dissection buffer: Hank's balanced salt solution (HBSS, no calcium, no magnesium, no phenol red; Life Technologies) with 1 mM HEPES (Life Technologies) and 100 U/mL penicillin and 100 μg/mL streptomycin (Life Technologies). Adjust pH to 7.4, filter (0.22 μm) and store at 4 °C.
5. Hemocytometer (Malassez™ or equivalent).
6. Seeding medium: On the day of seeding, prepare ~200 mL of DMEM (Dulbecco's modified eagle's medium, high glucose, pyruvate, Life Technologies) with 10 % horse serum (v/v, heat inactivated, Life Technologies).
7. Trypsin solution: 0.25 % trypsin-EDTA (Life Technologies) containing 15 mM HEPES and 33 mM D-glucose. Ready to use aliquots of 3 mL can be stored for weeks at −20 °C.
8. Trypsin inhibitor solution: Dissolve soybean trypsin inhibitor powder (Sigma) in the dissection buffer to a final concentration of 2 mg/mL. Ready to use aliquots of 3 mL can be stored for weeks at −20 °C.
9. Neuron culture media: Supplement ~250 mL of Neurobasal® media (Life Technologies) with 2 % B27® (v/v, serum-free

supplement, Life Technologies), 2 mM glutamine, 1 mM sodium pyruvate.

10. Cytosine-β-D-arabinofuranoside hydrochloride (AraC, Sigma) solution: Dissolve AraC in sterile distilled water to obtain a stock solution at 10 mM. Aliquot and store at −20 °C.

2.2 Release of Exosomes by Mature Cortical Neurons

1. K5 solution: 5 mM KCl, 1.8 mM $CaCl_2$, 0.8 mM $MgSO_4$, 110 mM NaCl, 26 mM $NaHCO_3$, 1 mM NaH_2PO_4, 0.7% (w/v) D-glucose and 15 mM HEPES. Adjust pH to 7.4. Filter (0.22 μm) and keep at 4 °C (*see* **Note 1**).
2. Bicuculline stock solution: Dissolve bicuculline methiodide powder (Sigma) in pure distilled water at 5 mM, aliquot, and store at −80 °C.
3. Stimulating solution: On the day of experiment, prepare K5 solution with 40 μM of bicuculline. Prepare 80 mL for the stimulation of twenty 100 mm dishes of neurons (52×10^6 neurons).

2.3 Exosome Purification

1. 50 mL tubes that can withstand 20,000 × *g* centrifugation (e.g., Falcon® tubes).
2. Centrifuge allowing 20,000 × *g* run (Sorvall Sc60 or equivalent).
3. Fixed-angle rotor allowing 20,000 × *g* centrifugation (F13-14X50CY FIBERLite® Thermo Scientific or equivalent).
4. Swinging rotor (SW41 and SW32 from Beckman Coulter or equivalent).
5. Ultracentrifuge (Optima L-90K or equivalent).
6. Ultracentrifugation tubes in polyallomer or Ultra-Clear™ for SW41 (maximum 12 mL) and SW32 (maximum 35 mL) swinging rotors.
7. Gradient maker connected to a peristaltic pump and a glass capillary tube.
8. Dulbecco's phosphate-buffered saline (DPBS, no calcium, no magnesium, Life Technologies).
9. 8% and 60% sucrose solutions (w/v) in distilled water containing 3 mM imidazole. 6 mL of each solution are required for one gradient. Adjust to pH 7.4. Filter solutions through a 0.22 μm sterile filter. Store at 4 °C.
10. Refractometer 0–72% (Euromex or equivalent).
11. Washing solution: 3 mM imidazole in distilled water. Adjust to pH 7.4. Store at 4 °C. ~100 mL are required for one gradient.
12. Hydrophilic syringe filter unit, 0.22 μm, PVDF (Millex-GV Durapore® Millipore or equivalent).

3 Methods

Pre-warm all media and solutions to 37 °C except when specified. Precool all centrifuges to 4 °C except when another temperature is indicated.

Four 100 mm dishes (1×10^7 neurons) are required to complete purification of exosomes at **step 12** of Subheading 3.2 ("exosome pellet"). Twenty 100 mm dishes (52×10^6 neurons) are needed when exosomes need to be purified on a linear gradient of sucrose (**steps 1–22** of Subheading 3.2).

3.1 Primary Culture of Cortical Neurons from 18-Day-Old Rat Embryos

1. Coating of dishes: Add 4 mL of the PDL solution per 100 mm dish, i.e., 2.6 μg PDL/cm² (*see* **Note 2**). Leave dishes for at least 4 h before seeding in a humidified incubator at 37 °C. Rinse dishes three times with sterile water. Finally, add 8 mL seeding medium per dish and incubate at 37 °C in a humidified incubator.
2. Dissection of rat embryo brains (*see* **Note 3**): Isolate embryos and cover them with dissection solution. Dissect out cortices from rodent brains (*see* **Note 4**) and pool them in 15 mL tubes filled with dissection solution.
3. Dissociation of cortical neurons by trypsin: Let cortices sediment in the 15 mL tube. Remove supernatant and add 3 mL of pre-warmed trypsin solution. Place the tube horizontally in the incubator at 37 °C, 5% CO_2 for 15 min. Turn it back to vertical and let cortices sediment. Remove trypsin solution. Add 5 mL of dissection solution to wash the cortices. Let the cortices sediment. Remove the supernatant. Repeat the wash once more. Add 3 mL of trypsin inhibitor solution. Place the tube horizontally in the incubator at 37 °C for 2 min. Turn it back to vertical and let cortices sediment. Remove the trypsin inhibitor solution. Rinse twice with dissection solution. Finally, add 5 mL of dissection solution to the cleaned cortices.
4. Mechanical dissociation of cortical neurons: Aspirate and flush the cortices first with a 5 mL pipette, then with a 1 mL tip (~40 ups and downs) to obtain a homogeneous solution. Be careful to avoid bubbles.
5. Filter the homogenate through a 70 μm filter placed above a 50 mL tube to remove debris. Centrifuge the cells for 5 min at room temperature at $100 \times g$. Gently remove the supernatant, then add 5 mL of seeding medium to the pellet and slowly homogenate the cell suspension. Keep the cell suspension at 37 °C in a water bath.
6. Counting neurons: Mix 10 μL of the cell suspension with 170 μL HBSS solution and 20 μL trypan blue. Load this cell dilution into a hemocytometer. Count only living cells (dead cells will be dyed blue) under the microscope (*see* **Note 5**).

7. Seeding cortical neurons: Plate neurons at a density of 3×10^4 cells/cm^2, i.e., 26×10^5 neurons per 100 mm coated dish. An example of counting and seeding is given in **Note 6**.
8. Primary cultures (*see* **Note 7**): After 2 h, replace 75% seeding medium by neuron culture medium. Repeat this again immediately. This can be done up to 24 h after plating neurons (*see* **Note 8**).
9. After 4 days in vitro (DIV), add 20% complete culture medium supplemented with 5 μM AraC (final concentration) per dish. AraC is used to stop glial cells proliferation (*see* **Note 9**).
10. At 11 DIV, add 10% pre-warmed culture medium to each dish.
11. Cortical neurons can be used at DIV14-15.

3.2 Glutamatergic Synapse Activation by Bicuculline Incubation and Exosome Purification

1. Wash the dishes (*see* **Note 10**) by removing 80% of the culture medium and gently adding 4 mL of K5 media per dish. Repeat once.
2. Add 4 mL per dish of bicuculline-containing solution (*see* **Note 11**).
3. Leave dishes 15 min at 37 °C in the incubator.
4. Harvest and pool supernatants of bicuculline-treated cells and keep all tubes on ice during exosome purification.
5. Centrifuge supernatants for 10 min at $2{,}000 \times g$ at 4 °C (*see* **Note 12**).
6. Transfer each supernatant to 50 mL tubes.
7. Centrifuge for 30 min at $20{,}000 \times g$ at 4 °C (*see* **Note 13**).
8. Filter supernatants through a 0.22 μm filter and transfer them to ultracentrifuge tubes (*see* **Note 14**). Tubes for a SW41 rotor have to be filled up to 10 mL. Tubes for a SW32 rotor have to be filled up to 30 mL.
9. Weigh and equilibrate the tubes with PBS.
10. Centrifuge for 90 min at $100{,}000 \times g$ at 4 °C (*see* **Note 15**).
11. Gently remove the supernatant and invert the tubes on an absorbent paper. The pellet corresponds to the "exosome pellet" (*see* **Note 16**).
12. You can stop the purification at this step and resuspend the exosome pellet in SDS-sample buffer for Western blot analysis (Fig. 1a, b).
13. For further purification of exosomes on a sucrose gradient, gently resuspend the exosome pellet in 300 μL 8% sucrose solution (*see* **Notes 17** and **18**).
14. Make a continuous gradient from 8 to 60% sucrose in a SW41 tube (*see* **Note 18**).

15. Gently add the 300 μL exosome sample from **step 13** to the top of the gradient (*see* **Note 19**).
16. Centrifuge the gradient 18 h at 210,000 × *g*, at 4 °C.
17. Collect 1 mL fractions starting from the top of the gradient and transfer each fraction to a new SW41 ultracentrifuge tube. Gently shake each fraction to homogenize the sucrose concentration.
18. Place 10 μL of each fraction in the refractometer to measure their sucrose density (*see* **Note 20**).
19. Add 10 mL of washing solution to each SW41 tube containing the fractions and mix.
20. Ultracentrifuge all fractions for 2 h at 210,000 × *g* at 4 °C (*see* **Note 21**). Tubes can be kept on ice before centrifugation.
21. Discard the supernatant and resuspend each pellet in the required buffer (e.g., SDS-sample buffer for Western blot analysis). Analyses of the exosome pellet from **step 12** and of fractions of a typical gradient are shown in Fig. 1.

4 Notes

1. An improvement of these protocols is to replace K5 solution by Neurobasal® medium (Life Technologies) to harvest exosomes. This avoids any shock caused by a change in media composition.
2. The PDL concentration used for coating is critical for the viability of the culture. Take care to homogenize the solution before use and do not leave the coated dishes more than 48 h in the incubator.
3. All the instruments (such as forceps and scissors) must be sterile. In order to avoid neuronal mortality, the time between harvesting of embryos and the seeding of cells should not exceed 2 h. Generally, 9 embryos are enough for twenty 100 mm dishes of neurons (52×10^6).
4. Dissection of pup brains (similar to E18 embryos brain) is described in [21]. Briefly, under a sterile hood, decapitate each embryo and transfer the heads to a dish filled with pre-warmed dissection solution. Insert the tips of fine scissors into the aperture made by the section of the vertebral column and cut forward, along the skull dorsolaterally, taking care not to damage the brain. Remove the top of the skull to uncover the brain. Slide the closed tips of the forceps underneath the brain to push it out from the skull. Transfer it to a new dish filled with clean dissection solution and then work under a binocular

microscope for the next steps which need fine forceps. Separate the two hemispheres, remove meninges, then cleave off the olfactory bulbs, brain stem, and the cerebellum. Finally, dissect out the hippocampus to obtain clean cortices.

5. Counting is a critical step because the seeding density determines the viability of the primary culture.
6. From ~12 embryos, we generally obtain ~14×10^7 living neurons in the 5 mL cell suspension. After diluting ten times in seeding medium, we distribute ~950 μL of the cell suspension per 100 mm dish to obtain a seeding density of 2.6×10^6 neurons per 100 mm dish (3×10^4 cells/cm^2).
7. Primary cultures are highly sensitive to changes in pH and temperature. Therefore, avoid opening and closing the incubator as much as possible over the whole culture period. Keep an incubator dedicated to long-term neuronal culture, if possible. Do not leave neurons outside the incubators for too long.
8. Never replace 100% of the culture media to avoid drying the cells. Aspirate only 6 mL of seeding media and replace it gently without flushing by 6 mL of the culture medium.
9. Primary cultures can be grown without AraC. In this case, the cultures contain a mix of neurons and glial cells. They are less fragile and can be kept alive longer than when grown with AraC. However, as described in the introduction, exosomes are secreted by glial cells present in the culture (astrocytes, oligodendrocytes, and microglia). We use an antibody against L1-CAM to differentiate exosomes released by neurons from exosomes released by glial cells since L1-CAM is exclusively expressed by neurons (see Fig. 1b).
10. Always check the viability of each dish under a microscope before use. Do not use cultures if the neurons are not refringent.
11. As a control, two dishes are treated in the same conditions with K5 without bicuculline. In our hands, only few exosomes are released without bicuculline treatment. See Fig. 1a, b to compare the two conditions.
12. This step removes dead cells and nuclei.
13. This step removes large debris and microvesicles.
14. This step further cleans the supernatants, in particular removes vesicles bigger than 200 nm.
15. Different times and speeds of ultracentrifugation can be found in the literature to pellet nanovesicles. Run for at least 60 min at $100{,}000 \times g$ (calculation based on the speed at the bottom of the tube). However, many ultracentrifuges include the acceleration and/or deceleration phases in their timing. It is thus

recommended to increase the length of the run. Running too slow can lead to a loss of material. Running too fast (more than 200,000 × *g*) and too long (more than 2 h) can lead to contamination of the pellet by lighter particles such as endoplasmic reticulum or mitochondria-derived vesicles.

16. Exosome pellets purified from neuronal supernatants are not visible. White opaque material is not relevant to vesicle pellets but to contaminating debris. Nanovesicle pellets are known to adhere quite strongly to plastic tubes, so prolonged but gently flushing (at least 100 aspirations with a pipette) is needed to resuspend pellets correctly. Flush only the bottom but not the walls of the tube and avoid bubbles.
17. Harvest 30 μL from the 300 μL and mix directly with a concentrated SDS-sample buffer ("10% input"; Fig. 1c).
18. Various protocols for making sucrose gradient can be found in the literature [22]. Before pouring the gradient, mark the ultracentrifuge tube at the 10 mL level. In the gradient former, add 5.5 mL of 60% sucrose solution in chamber 1 and 5.5 mL 60% sucrose solution in chamber 2 which is linked by tubing to the pump. Avoid bubbles in the tubules between the two chambers. Place the magnetic bar in chamber 2, switch on the stirring, and check the efficiency and regularity of the stirring speed. Place the capillary at the bottom of the ultracentrifuge tube. Switch on the pump at a speed of ~1 mL/min and immediately open the clamp between the two chambers. Stop the pump when the gradient volume reaches the 10 mL mark. Delicately remove the capillary from the tube.
19. To balance the rotor, fill a second ultracentrifuge tube with 5.3 mL of 8% sucrose solution and 5 mL of 60% sucrose solution. Equilibrate the tubes with the 8% sucrose solution.
20. Check the linearity of the gradient by tracing the sucrose density as a function of the fraction.
21. The purpose of this ultracentrifugation step is to concentrate vesicles and wash out sucrose from each fraction. The speed and length of centrifugation is thus faster and longer than the first ultracentrifugation in **step** 7 without risk of contamination.

Acknowledgment

K.L. was supported by "Fondation Plan Alzheimer." C.J. and M.C. were supported by the "Ministère de l'Enseignement Supérieur et de la Recherche." This work was funded by INSERM, Université Grenoble Alpes, and ANR (08-Blanc-0271 to R.S.).

References

1. Thery C, Ostrowski M, Segura E (2009) Membrane vesicles as conveyors of immune responses. Nat Rev Immunol 9:581–593
2. Keller S, Sanderson MP, Stoeck A, Altevogt P (2006) Exosomes: from biogenesis and secretion to biological function. Immunol Lett 107: 102–108
3. Kalra H, Simpson RJ, Ji H et al (2012) Vesiclepedia: a compendium for extracellular vesicles with continuous community annotation. Plos Biol 10(12):e1001450
4. Lasser C, Alikhani VS, Ekstrom K et al (2011) Human saliva, plasma and breast milk exosomes contain RNA: uptake by macrophages. J Transl Med 9:9
5. Vella LJ, Greenwood DL, Cappai R et al (2008) Enrichment of prion protein in exosomes derived from ovine cerebral spinal fluid. Vet Immunol Immunopathol 124:385–393
6. Raposo G, Nijman HW, Stoorvogel W et al (1996) B lymphocytes secrete antigen-presenting vesicles. J Exp Med 183:1161–1172
7. Thery C, Zitvogel L, Amigorena S (2002) Exosomes: composition, biogenesis and function. Nat Rev Immunol 2:569–579
8. Wang G, Dinkins M, He Q et al (2012) Astrocytes secrete exosomes enriched with proapoptotic ceramide and prostate apoptosis response 4 (PAR-4) potential mechanism of apoptosis induction in Alzheimer disease (AD). J Biol Chem 287:21384–21395
9. Potolicchio I, Carven GJ, Xu X et al (2005) Proteomic analysis of microglia-derived exosomes: metabolic role of the aminopeptidase CD13 in neuropeptide catabolism. J Immunol 175:2237–2243
10. Faure J, Lachenal G, Court M et al (2006) Exosomes are released by cultured cortical neurones. Mol Cell Neurosci 31:642–648
11. Fevrier B, Vilette D, Laude H, Raposo G (2005) Exosomes: a buble ride of prions? Traffic 6:10–17
12. Smalheiser NR (2007) Exosomal transfer of proteins and RNAs at synapses in the nervous system. Biol Direct 2:35
13. Kramer-Albers EM, Bretz N, Tenzer S et al (2007) Oligodendrocytes secrete exosomes containing major myelin and stress-protective proteins: trophic support for axons? Proteomics Clin Appl 1:1446–1461
14. Lopez-Verrilli MA, Picou F, Court FA (2013) Schwann cell-derived exosomes enhance axonal regeneration in the peripheral nervous system. Glia 61:1795–1806
15. Laulagnier K, Motta C, Hamdi S et al (2004) Mast cell- and dendritic cell-derived exosomes display a specific lipid composition and an unsual membrane organization. Biochem J 380:161–171
16. Savina A, Vidal M, Colombo MI (2002) The exosome pathway in K562 cells is regulated by Rab11. J Cell Sci 115:2505–2515
17. Vincent-Schneider H, Stumptner-Cuvelette P, Lankar D et al (2002) Exosomes bearing HLA-DR1 molecules need dendritic cells to efficiently stimulate specific T cells. Int Immunol 14:713–722
18. Bading H, Greenberg ME (1991) Stimulation of protein tyrosine phosphorylation by NMDA receptor activation. Science 253: 912–914
19. Ichikawa M, Muramoto K, Kobayashi K et al (1993) Formation and maturation of synapses in primary cultures or rat cerebral cortical cells: an electron microscopic study. Neurosci Res 16:95–103
20. Lachenal G, Pernet-Gallay K, Chivet M et al (2011) Release of exosomes from differentiated neurons and its regulation by synaptic glutamatergic activity. Mol Cell Neurosci 46: 409–418
21. Beaudoin GM 3rd, Lee SH, Singh D et al (2012) Culturing pyramidal neurons from the early postnatal mouse hippocampus and cortex. Nat Protoc 7:1741–1754
22. Marks MS (2001) Determination of molecular size by zonal sedimentation analysis on sucrose density gradients. Curr Protoc Cell Biol Chapter 5:Unit 3

Chapter 10

A Method for Isolation of Extracellular Vesicles and Characterization of Exosomes from Brain Extracellular Space

Rocío Pérez-González*, Sebastien A. Gauthier*, Asok Kumar, Mitsuo Saito, Mariko Saito, and Efrat Levy

Abstract

Extracellular vesicles (EV), including exosomes, secreted vesicles of endocytic origin, and microvesicles derived from the plasma membrane, have been widely isolated and characterized from conditioned culture media and bodily fluids. The difficulty in isolating EV from tissues, however, has hindered their study in vivo. Here, we describe a novel method designed to isolate EV and characterize exosomes from the extracellular space of brain tissues. The purification of EV is achieved by gentle dissociation of the tissue to free the brain extracellular space, followed by sequential low-speed centrifugations, filtration, and ultracentrifugations. To further purify EV from other extracellular components, they are separated on a sucrose step gradient. Characterization of the sucrose step gradient fractions by electron microscopy demonstrates that this method yields pure EV preparations free of large vesicles, subcellular organelles, or debris. The level of EV secretion and content are determined by assays for acetylcholinesterase activity and total protein estimation, and exosomal identification and protein content are analyzed by Western blot and immuno-electron microscopy. Additionally, we present here a method to delipidate EV in order to improve the resolution of downstream electrophoretic analysis of EV proteins.

Key words Extracellular vesicles, Exosomes, Brain, Extracellular space, Differential ultracentrifugation, Sucrose step gradient, Delipidation

1 Introduction

Although described more than 40 years ago, the release of membrane-enclosed vesicles by cells into the extracellular space has been the subject of increasing interest in the past few years. EV are phospholipid bilayer membrane-enclosed secreted vesicles that contain lipids, proteins, and RNA (mRNA and miRNA). Exosomes are one subtype of membrane-enclosed secreted vesicles with a size ranging from 20 to 150 nm. They form intracellularly by inward

*Both authors contributed equally to this work.

Andrew F. Hill (ed.), *Exosomes and Microvesicles: Methods and Protocols*, Methods in Molecular Biology, vol. 1545, DOI 10.1007/978-1-4939-6728-5_10, © Springer Science+Business Media LLC 2017

budding of the endosomal membrane, leading to the formation of multivesicular bodies (MVBs). The internal vesicles are released as exosomes when MVBs are fused with the plasma membrane. While most published studies refer to isolated vesicles as exosomes, the methodology used in these studies most likely results in the isolation of EV of multiple types and multiple origins. EV have been purified from cell culture conditioned media [1] and bodily fluids, such as cerebrospinal fluid [2], plasma [3], and urine [4], by different methods. The most commonly used protocol for EV isolation is differential ultracentrifugation, which is usually followed by an extra purification step, such as a sucrose gradient centrifugation [5]. Several alternative methods have been recently introduced and utilized for isolation and purification of exosomes including antibody-coated magnetic beads [6], microfluidic devices [7], precipitation technologies [8], and microfiltration approaches [9]. The difficulty in isolation of EV from tissues, however, has delayed their study in vivo. We have developed a protocol to isolate EV from the extracellular space of human and murine brain tissues. Electron microscopy (EM) imaging of EV shows that this method yields a pure EV preparation free of large vesicles, subcellular organelles, and debris (Fig. 1). Further, EM analysis revealed no difference between EV isolated from brain tissues frozen for several years at −80 °C and EV from freshly isolated mouse brains [10]. Therefore, this procedure can be applied to postmortem human material that has been frozen for long periods of time. The first step of the isolation procedure involves mild treatment of

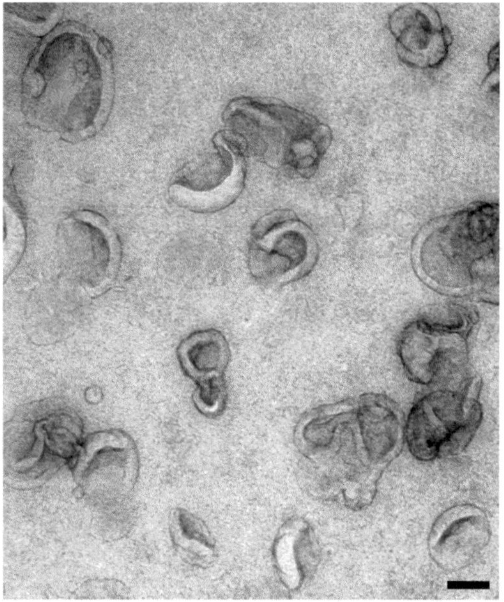

Fig. 1 EV isolated from frozen human brain. Wide-field EM imaging shows multiple human brain EV of sizes ranging from 20 to 150 nm and no other structures or cellular debris. *Scale bar* 100 nm

tissue with papain in order to dissociate the tissue and free the extracellular space, containing brain EV. This procedure is widely used for harvesting viable primary neurons from fresh brain tissue [11]. The mild papain treatment does not trigger cell lysis and thus prevents the contamination of the extracellular fluid with intracellular organelles and vesicles. Low-speed centrifugation to remove cells and debris is followed by filtration of the brain extracellular material to eliminate from the samples vesicles larger than 200 nm and protein aggregates. EV are isolated from the filtrate by differential ultracentrifugation as described previously [5], including a final sucrose step gradient centrifugation. The centrifugation of the sucrose step gradient at high-speed results in migration of EV through the gradient and in their settlement at the interphase between two adjacent sucrose layers whose densities flank the EV density. EV equilibrate with sucrose solution at a density ranging from 1.09 to 1.17 g/ml. Thus, the final purification step eliminates possible contamination with other types of extracellular vesicles, proteins nonspecifically associated with EV, and large protein aggregates, which do not have the EV density. The sucrose step gradient centrifugation resolves a sample into seven different fractions, from the top lightest *a* fraction to the bottom *g* heaviest fraction. Western blot analysis for the exosomal marker Alix [12] (Fig. 2a) and EM imaging (Fig. 2b) have shown the presence of exosomes only in fractions *b*, *c*, and *d*. This purification method yields large enough amounts of exosomes allowing for multiple downstream analyses. An adult/juvenile mouse hemibrain usually yields from 200 to 300 μg of total EV proteins. Similar yields are obtained from 0.2 g of human brain tissue.

New methods to isolate EV from cell media and bodily fluids are being commercialized and are finding their way to common

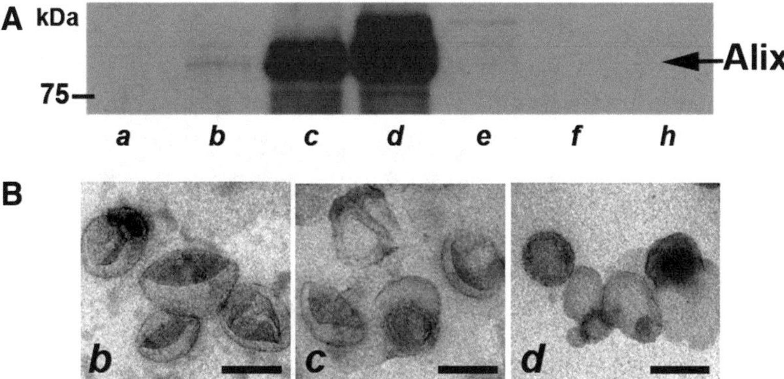

Fig. 2 Characterization of exosomes isolated from a murine brain. **A**. Western blot analysis demonstrating the presence of the exosomal marker Alix in fractions *b*, *c*, and *d*, collected from the sucrose step gradient column. **B**. EM imaging showing mouse brain EV of sizes ranging from 20 to 150 nm in fractions *b*, *c*, and *d*. *Scale bar* 100 nm

practice. In an attempt to speed up the isolation procedure, we tested one of the kits designed for EV isolation from cell medium to isolate EV from brain tissue. In order to do that, we first followed our protocol to remove brain cells and generate a cell medium-like product that was processed according to the kit manufacturer's protocol. Unfortunately, the yield of the EV obtained using the commercially available kit was low and the preparations were contaminated with cellular debris, as observed by EM.

EV, especially exosomes, are limited by a lipid bilayer enriched in cholesterol and sphingolipids (reviewed in [13]). When EV proteins are analyzed by gel electrophoresis, lipids interfere with the run, especially of low-molecular-weight protein bands (Fig. 3). The removal of lipids from the EV before electrophoretic analysis improves the resolution of the bands on electrophoretic gels (Fig. 3). Therefore, we describe here a protocol to delipidate EV that is based on the method of lipid extraction described by Saito et al. [14] and Matyash et al. [15]. While the lipid profile of brain EV can be studied from the soluble fraction, the protein content can be analyzed from the precipitate.

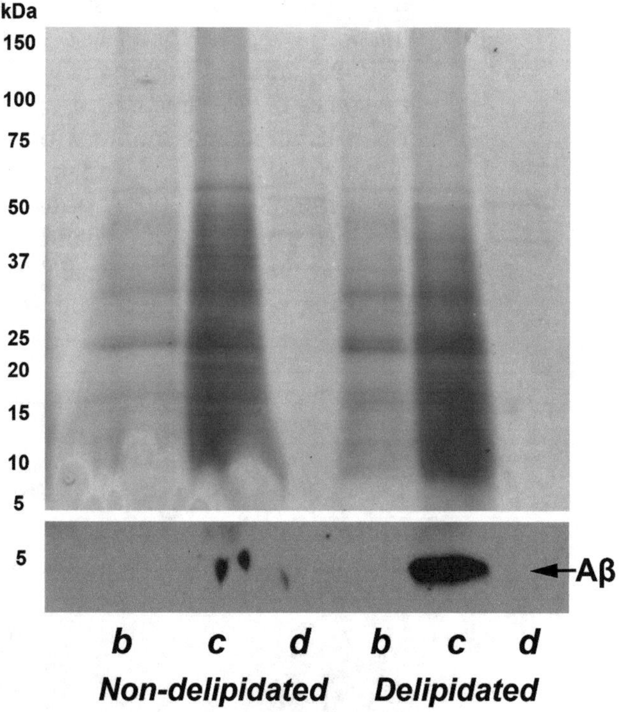

Fig. 3 Improved resolution of EV protein bands in gel electrophoresis by delipidation. The representative 4–20 % Tris–HCl gel stained with Coomassie blue (*upper panel*) shows non-delipidated and delipidated EV proteins in fractions *b*, *c*, and *d*. Western blot of the same proteins with 4G8 antibody that reacts with amyloid β (Aβ) (*lower panel*) shows EV Aβ in the brain of a mouse model of β amyloidosis, Tg2576, at 24 months of age, before and after delipidation

The protocol described here requires rigor and precise performance and deviation from some of its details, such as the use of the wrong rotor for ultracentrifugation, would prevent success.

2 Materials

2.1 Supplies/Reagents Used for Isolation of EV from Brain Tissues

1. Freshly removed or previously frozen at −80 °C murine or human brain tissue. Brain tissue sample is one-mouse hemi-brain or 0.2-g human tissues.
2. 70% ethanol.
3. Sterile tools for dissecting brain tissues (including scissors, tweezers/forceps, and razor blade/scalpel) and a dissecting pad.
4. Dry ice to flash-freeze tissues.
5. Hibernate A (BrainBits, LLC, HA) kept at 4 °C.
6. Papain vials (Worthington Biochemicals, LK003178). Freshly prepare a solution of 20 units/ml papain in Hibernate A and keep at 37 °C (3.5 ml per sample).
7. Plastic transfer pipettes.
8. Protease inhibitors: PMSF (Phenylmethylsulfonyl fluoride, Sigma P7626) and the cocktail LAP (Leupeptin, Sigma L2884; Antipain, Sigma A6191; Pepstatin A, Sigma P4265).
9. 15-ml conical tubes.
10. 10-ml plastic pipettes.
11. 50-ml conical tubes.
12. 1× phosphate-buffered saline (PBS) without calcium and magnesium at pH 7.4 kept at 4 °C.
13. 70-ml high-speed centrifugation polycarbonate tubes (Beckman Coulter, 355622).
14. 0.2-μm filter adapters with surfactant-free cellulose acetate membrane (Corning, 431219) and 20-ml plunge syringes.
15. 1.5-ml microfuge tubes.

2.2 Supplies/Reagents Used for Purification of EV on a Sucrose Step Gradient

1. N-2-Hydroxyethylpiperazine-N′-2-ethane sulfonic acid (HEPES), pH 7.5, 1 M.
2. Sucrose (Reagent, A.C.S.). Freshly prepare 2 ml each of 0.25, 0.6, 0.95, 1.3, 1.65, and 2 M sucrose solutions in 20 mM HEPES (per sample).
3. PBS kept at 4 °C.
4. 13-ml high-speed ultra-clear centrifuge tubes (Beckman Coulter, 344060).
5. 10-ml high-speed centrifugation tubes (Beckman Coulter, 355647).
6. 0.5-ml microfuge tubes.

2.3 Supplies/Reagents Used for Characterization of Brain EV

2.3.1 EM Imaging

1. 20% paraformaldehyde (PFA) (Reagent, A.C.S.).
2. PBS kept at 4 °C.
3. 1% glutaraldehyde (EM grade distillation purified).
4. Uranyl-oxalate, pH 7.0.
5. Methyl cellulose (25 cP, Sigma M-6385).
6. Uranyl acetate (UA) (Reagent, A.C.S.), pH 4.0. Prepare nine parts 2% methyl cellulose and one part 4% UA mixed just before use.
7. 300 mesh Formvar carbon-coated EM grids (Electron Microscopy Sciences, FCF300).
8. Parafilm.
9. Clean forceps.
10. Glass dish.
11. Stainless steel loops.
12. Whatman number 1 filter paper.

2.3.2 Acetylcholinesterase (AChE) Activity Assay

1. Acetylthiocholine iodide (Sigma-Aldrich, A5751).
2. 5,5-Dithiobis(2-nitrobenzoic acid) (DNTB) (Sigma-Aldrich, D8130).
3. Freshly prepare AChE activity assay working solution containing 1.25 mM acetylthiocholine iodide and 0.1 mM DNTB in 0.1 M phosphate buffer (PB) at pH 8.0.
4. Flat bottom 96-well plate and adhesive cover.

2.3.3 EV Protein Content Analysis

1. 2× RIPA lysis buffer (2% Triton-X, 2% sodium deoxycholate, 0.2% SDS, 300 mM NaCl, 100 mM Tris–HCl pH 7.4, and 1 mM EDTA in double distilled water) freshly supplemented with protease inhibitors (2 mM PMSF and 0.02% v/v LAP).
2. BCA protein assay kit (Pierce™, 23225).
3. Flat bottom 96-well plate and adhesive cover.

2.3.4 Western Blot Analysis

1. 6× SDS loading buffer (0.375 M Tris–HCl pH 6.8, 12% SDS, 50% glycerol, 30% β-mercaptoethanol, 0.06% bromophenol blue).
2. 4-20% Tris–HCl electrophoresis gel.
3. 5× electrophoretic running buffer/6× transfer buffer (0.125 M Tris base, 1.25 M glycine). When diluting to 1×, add SDS to 0.1% for electrophoretic running buffer, or methanol to 20% for transfer buffer.
4. 0.45 μm polyvinylidene fluoride membrane (PVDF) (Immobilon).
5. TBST (10 mM Tris, 150 mM sodium chloride, pH 7.5, 0.1% Tween-20).

6. Blocking buffer (5 % milk, Bio Rad) in TBST.
7. Primary antibodies diluted in TBST, such as anti-Alix (EMD Millipore, ABC40) and antibody 4G8 (BioLegend, 39220).
8. Secondary antibodies diluted in TBST, such as anti-mouse IgG (GE Healthcare Life Sciences) and anti-rabbit IgG antibodies (GE Healthcare Life Sciences).
9. ECL Western Blotting Substrate (Thermo Scientific) and SuperSignal West Femto Maximum Sensitivity Substrate (Thermo Scientific).
10. Reflection autoradiography film.

2.4 Supplies/Reagents Used for Delipidation of EV

1. Dry ice to flash-freeze the EV.
2. Filter paper (EMS Lens tissue, Electron Microscopy Sciences) and rubber bands to attach the filter paper to the tubes.
3. Methyl tertially butyl ether (MTBE) (HPLC grade)/methanol (HPLC grade) (1:1) to solubilize EV lipids.
4. 1× RIPA lysis buffer prepared by diluting 2× RIPA buffer with protease inhibitors in PBS.

3 Methods

The following protocols describe the isolation of EV from mouse hemibrain tissue (Subheading 3.1), the purification of EV on a sucrose step gradient column (Subheading 3.2), different assays to characterize the EV in general and exosomes in particular (Subheading 3.3), and a method to delipidate EV (Subheading 3.4). All the procedures can be applied to 0.2 g of human brain tissue. Brain EV levels can be quantified by measuring either EV AChE activity (Subheading 3.3.2) or EV protein content (Subheading 3.3.3), calculating the total level of either AChE activity or protein content in the EV isolated from the hemibrain and standardization to the hemibrain weight. The level of exosome secretion into the brain extracellular space can be calculated from Western blots of exosomal protein markers (e.g., Alix, Tsg101, or CD63) (Subheading 3.3.4) by quantification of the protein bands and standardization to the hemibrain weight. Specific protein content within exosomes can be calculated by standardizing the levels of that specific protein to an exosomal marker such as Alix quantified from a Western blot.

3.1 Isolation of EV from Brain Tissues

1. Sacrifice a mouse by a method approved by your animal care committee.
2. Spray the fur of the mouse around the head/neck region with 70% ethanol and decapitate the animal.

3. Remove the mouse brain from the skull and place it on a cool dissecting pad. Make sure the brain is clean of hair and skin bits.
4. Cut out the cerebellum and the olfactory bulbs. Separate and weight the two hemibrains. The hemibrain used for EV isolation can be flash frozen and processed later. Always use the same hemibrain for the same application, for example, always use the right hemibrain for EV isolation and the left one for preparing brain homogenate.
5. Finely mince the hemibrain (fresh or frozen) used for EV isolation with a razor blade/scalpel in a few drops of a 20 units/ml papain solution in Hibernate A kept at 37 °C (*see* **Note 1**).
6. Transfer the brain tissue with a plastic transfer pipette into a 15-ml conical tube containing 3.5 ml of 20 units/ml papain in Hibernate A kept at 37 °C.
7. Incubate for 20 min at 37 °C to dissociate the tissue and free the extracellular space. Gently shake (not vortex) the tube every 5 min. In the meantime repeat the procedure from **step 1** for the next mouse (*see* **Note 2**).
8. Stop the enzymatic papain reaction by adding 6.5 ml of ice cold Hibernate A supplemented with protease inhibitors (2 mM PMSF and 0.02% v/v LAP).
9. Pass the tissue in solution through the tip of a 10-ml plastic pipette (15–20 times) to further loosen up the tissue until the solution is homogenous.
10. Centrifuge the filtrate at 300 ×*g* for 10 min at 4 °C to discard brain cells.
11. Collect the supernatant and centrifuge at 2000 ×*g* for 10 min at 4 °C to discard large cell debris.
12. Transfer the supernatant into a 70-ml high-speed centrifugation polycarbonate tube.
13. Add ice cold PBS to bring the volume up to 60 ml (*see* **Notes 3** and **4**).
14. Ultracentrifuge the tubes for 30 min at 10,000 ×*g* at 4 °C using a fixed-angle rotor (e.g., Beckman 45Ti rotor for 70-ml tubes in a Sorvall WX ultra 80 centrifuge) (*see* **Note 5**) to discard small cell debris.
15. Collect carefully the supernatant leaving the pellet undisturbed (*see* **Note 6**), and pass it through a 0.2-μm syringe filter adapter into another 70-ml polycarbonate tube. This filtration step eliminates larger vesicles from the EV preparations.
16. Add ice cold PBS to bring the volume up to 60 ml (*see* **Notes 3** and **4**).
17. Ultracentrifuge the tubes for 70 min (10 min to get up to speed and 1 h at max speed) at 100,000 ×*g* at 4 °C using a

fixed-angle rotor (e.g., Beckman 45Ti rotor for 70-ml tubes in a Sorvall WX ultra 80 centrifuge) (*see* **Note 5**) to pellet the EV.

18. Discard the supernatant without disturbing the EV pellet. The supernatant can be poured off.
19. Gently resuspend the pellet in 1 ml of PBS kept at 4 °C, and fill the tube up to a volume of 60 ml with cold PBS (*see* **Notes 3** and **4**). This step will wash the EV in a large volume of PBS to eliminate contaminating proteins.
20. Ultracentrifuge the tubes for 70 min at 100,000 × *g* at 4 °C using a fixed-angle rotor (e.g., Beckman 45Ti rotor for 70-ml tubes in a Sorvall WX ultra 80 centrifuge) (*see* **Note 5**).
21. Discard the supernatant without disturbing the EV pellet. The supernatant can be poured off. Invert the tube and take out the leftover supernatant with a micropipette. Keep the pellet at 4 °C.
22. Resuspend the EV pellet in 180 μl of PBS: 90 μl to resuspend most of the EV and 90 μl to wash the tube and collect the remaining EV (*see* **Note 7**), and transfer into 1.5-ml microfuge tube.

3.2 EV Purification on a Sucrose Step Gradient

1. Resuspend the EV pellet in 2 ml of the 0.95 M sucrose solution (or if the EV are already resuspended in a small volume of PBS, add enough 0.95 M sucrose to reach a volume of 2 ml).
2. Carefully lay 2 ml sucrose solutions on top of each other starting with the highest concentration (2.0 M sucrose) at the bottom of a 13-ml high-speed ultra-clear centrifuge tube. Insert the 0.95 M sucrose solution containing the EV in the step sucrose gradient on top of the 1.30 M sucrose layer (*see* **Notes 4** and **8**).
3. Ultracentrifuge the tubes at 200,000 × *g* for 16 h using a swinging-bucket rotor (e.g., Beckman SW40Ti swinging rotor for 13-ml soft tubes in a Sorvall WX ultra 80 centrifuge).
4. Collect 1 ml fraction from the top of the gradient and then collect 2 ml fractions flanking the interphase between two adjacent sucrose step layers into 10-ml high-speed centrifugation tubes (*see* **Note 9**).
5. Add ice cold PBS to the tubes to bring up the volume to 8 ml (*see* **Notes 4–5**).
6. Ultracentrifuge the tubes at 100,000 × *g* for 1 h using a fixed-angle rotor (e.g., Beckman MLA-80 rotor for 10-ml tubes in a Beckman Optima MAX-E high-speed centrifuge).
7. Discard the supernatant without disturbing the EV pellet. The supernatant can be poured off. Invert the tube and take out the leftover supernatant with a micropipette. Keep the pellet at 4 °C.

8. Resuspend the fraction pellets in 30 μl of PBS: 15 μl to resuspend the pellet and 15 μl to wash the tube and collect the remaining EV, and transfer into 0.5-ml microfuge tubes.

3.3 Characterization of Brain EV

This protocol follows the method previously described by Thery et al. [5].

3.3.1 EM Imaging

1. Fix 2 μl of the EV in 2% PFA (add 20% PFA to the EV in PBS) (*see* **Note 10**).
2. Deposit 5-μl fixed EV on Formvar carbon-coated EM grids. Cover and let the membranes adsorb for 20 min in a dry environment.
3. Put 100-μl drops of PBS on a sheet of Parafilm. Transfer each grid on top of a drop of PBS with clean forceps to wash.
4. Transfer each grid onto a 50-μl drop of 1% glutaraldehyde for 5 min.
5. Transfer onto a 100-μl drop of distilled water and let grids stand for 2 min. Repeat seven times for a total of eight water washes.
6. Transfer the grid onto a 50-μl drop of uranyl-oxalate solution, pH 7.0, for 5 min.
7. Transfer the grid onto a 50-μl drop of methyl cellulose-UA for 10 min on ice.
8. Remove the grids with stainless steel loops, and blot excess fluid by gently pushing the loop sideways on Whatman number 1 filter paper so that a thin film is left behind over the EV side of the grid.
9. Air-dry the grid 5–10 min while still on the loop.
10. Observe under the electron microscope at 80 kV.

3.3.2 AChE Activity Assay (See Note 11)

1. Pipette 18 μl of 0.1 M PB, pH 8.0, into a 96-well plate.
2. Resuspend 2 μl of the EV in the PB, pH 8.0.
3. Add 280 μl of the AChE assay working solution to the well and mix the plate thoroughly on a plate shaker for 30 s.
4. Cover the plate and incubate at 37 °C in the dark for 45 min.
5. Measure the absorbance at 412 nm on a plate reader.

3.3.3 EV Protein Estimation

EV protein content can be determined by using the BCA protein assay kit on the EV lysate diluted 1:5 in distilled water (2 μl lysate + 8 μl distilled water) (*see* **Note 12**):

1. EV lysates are prepared by adding an equal volume of 2× RIPA buffer freshly supplemented with protease inhibitors to the EV suspended in PBS.
2. Pipette10 μl of each standard or EV sample into a 96-well plate.

3. Add 200 μl of the BCA working reagent to each well and mix the plate thoroughly on a plate shaker for 30 s.
4. Cover the plate and incubate at 37 °C for 30 min.
5. Cool the plate to room temperature and measure the absorbance at or near 562 nm on a plate reader.

3.3.4 EV Protein Content Analysis (See Note 13)

1. Add 6× SDS loading buffer to the EV lysate. Equal volumes of EV lysates (5–20 μl) are loaded in each lane.
2. Boil the samples at 96 °C for 5 min.
3. Separate EV proteins in 4–20% Tris–HCl electrophoresis gel.
4. Electrophoretically transfer the proteins onto a PVDF membrane.
5. Block the membrane in blocking solution for 1 h at room temperature.
6. Incubate the membrane with primary antibody overnight at 4 °C.
7. Incubate the membrane with secondary antibody for 1–2 h at room temperature.
8. Incubate the membrane in chemiluminescent fluid for 5 min and visualize on reflection autoradiography film.

3.4 Delipidation of EV

1. Freeze the EV resuspended in PBS (obtained at the end of Subheading 3.2) (*see* **Note 14**) and lyophilize the samples overnight. For the lyophilization process, lids of microfuge tubes should be open and covered with filter paper. The filter paper will allow evaporation of water while preventing the loss of proteins.
2. Add 200 μl of MTBE/methanol (1:1) and mix thoroughly using a bath-type sonicator and a vortex mixer.
3. Centrifuge at 500 × *g* for 5 min at room temperature.
4. Remove the supernatant without disturbing the pellet containing the EV proteins (*see* **Note 15**).
5. Dry the pellet in a vacuum concentrator and resolubilize the EV proteins in 1× RIPA lysis buffer containing protease inhibitors. EV proteins can be analyzed by Western blotting as described above (Subheading 3.3.4).

4 Notes

1. Prepare the papain solution fresh to minimize loss of enzyme activity. Incubate for 10 min at 37 °C before use.
2. With this protocol, EV can be isolated from up to four fresh hemibrains. Dissect the hemibrain of the second mouse, while

the hemibrain of the first animal is incubated in the papain solution, and so on with the other samples. This results in a 10-min delay between each hemibrain. Process all the brain tissue samples one after the other and place them on ice until the first centrifugation step. Then centrifuge all samples together. Up to six samples can be analyzed when EV are isolated from frozen brain tissues (which results in a 5-min delay between samples).

3. Always fill high-speed centrifuge tubes up to at least ¾ of the total volume capacity.
4. Make sure that all high-speed centrifuge tubes have the same weight (to the 0.01 g).
5. For ultracentrifugation steps in a fixed-angle rotor, mark the side of the tube facing up in the centrifuge rotor to point out the location of the pellet.
6. Leave enough supernatant in the tube so that the pellet stays submerged and does not contaminate the supernatant containing the EV.
7. The EV resuspended in PBS can be characterized following the procedures described in Subheading 3.3. If no characterization of the EV is desired at this point, the EV pellet can be directly resuspended in 0.95 M sucrose, as explained in Subheading 3.2.
8. Fill soft high-speed centrifuge tubes to the top to prevent the tube from collapsing during high-speed centrifugation.
9. This protocol generates seven fractions counting the top 1 ml of the gradient (first fraction), five 2 ml fractions containing the material accumulated at the interphases between two consecutive sucrose layers, and the bottom 1 ml of the gradient (last fraction). Fractions are named from the top to the bottom as *a*, *b*, *c*, *d*, *e*, *f*, and *g*. Fractions *b*, *c*, and *d* contain EV (Fig. 2a).
10. Store the fixed samples at 4 °C until analysis by EM. Regular EM analysis can be performed on the EV preparations before and after the purification onto the sucrose column. However, immuno-EM analyses are only doable in preparations before column, most likely due to interference with sucrose.
11. The AChE activity assay is based on the Ellman assay previously described [16].
12. EV protein content estimation can also be performed after the delipidation of EV. However, it is recommended doing it before delipidation as the proteins are more easily solubilized in the lysis buffer, improving the accuracy of the BCA protein assay.
13. It is advisable to perform Western blot analysis on delipidated samples to improve the resolution of the bands (Fig. 3).

14. Before freezing the EV in PBS, save a small volume for EM analysis and AChE activity assay.
15. The supernatant containing the soluble fraction (EV lipids) can be used to study the lipid profile of EV.

Acknowledgments

This work was supported by the National Institutes of Health (AG017617) and the Alzheimer's Association (NIRG-14-316622).

References

1. Rajendran L, Honsho M, Zahn TR, Keller P, Geiger KD, Verkade P, Simons K (2006) Alzheimer's disease β-amyloid peptides are released in association with exosomes. Proc Natl Acad Sci U S A 103(30):11172–11177
2. Street JM, Barran PE, Mackay CL, Weidt S, Balmforth C, Walsh TS, Chalmers RT, Webb DJ, Dear JW (2012) Identification and proteomic profiling of exosomes in human cerebrospinal fluid. J Transl Med 10:5. doi:1479-5876-10-5
3. Muller L, Hong CS, Stolz DB, Watkins SC, Whiteside TL (2014) Isolation of biologically-active exosomes from human plasma. J Immunol Methods 411:55–65
4. Wang D, Sun W (2014) Urinary extracellular microvesicles: isolation methods and prospects for urinary proteome. Proteomics 14(16): 1922–1932
5. Thery C, Amigorena S, Raposo G, Clayton A (2006) Isolation and characterization of exosomes from cell culture supernatants and biological fluids. Curr Protoc Cell Biol Chapter 3:Unit 3 22
6. Oksvold MP, Neurauter A, Pedersen KW (2015) Magnetic bead-based isolation of exosomes. Methods Mol Biol 1218:465–481
7. Kanwar SS, Dunlay CJ, Simeone DM, Nagrath S (2014) Microfluidic device (ExoChip) for on-chip isolation, quantification and characterization of circulating exosomes. Lab Chip 14(11):1891–1900
8. Alvarez ML (2014) Isolation of urinary exosomes for RNA biomarker discovery using a simple, fast, and highly scalable method. Methods Mol Biol 1182:145–170
9. Merchant ML, Powell DW, Wilkey DW, Cummins TD, Deegens JK, Rood IM, McAfee KJ, Fleischer C, Klein E, Klein JB (2010) Microfiltration isolation of human urinary exosomes for characterization by MS. Proteomics Clin Appl 4(1):84–96
10. Perez-Gonzalez R, Gauthier SA, Kumar A, Levy E (2012) The exosome secretory pathway transports amyloid precursor protein carboxyl-terminal fragments from the cell into the brain extracellular space. J Biol Chem 287(51): 43108–43115
11. Tizon B, Ribe EM, Mi W, Troy CM, Levy E (2010) Cystatin C protects neuronal cells from amyloid-β-induced toxicity. J Alzheimers Dis 19(3):885–894
12. Colombo M, Moita C, van Niel G, Kowal J, Vigneron J, Benaroch P, Manel N, Moita LF, Thery C, Raposo G (2013) Analysis of ESCRT functions in exosome biogenesis, composition and secretion highlights the heterogeneity of extracellular vesicles. J Cell Sci 126(Pt 24): 5553–5565
13. Record M, Carayon K, Poirot M, Silvente-Poirot S (2014) Exosomes as new vesicular lipid transporters involved in cell-cell communication and various pathophysiologies. Biochim Biophys Acta 1841(1):108–120
14. Saito M, Chakraborty G, Shah R, Mao RF, Kumar A, Yang DS, Dobrenis K (2012) Elevation of GM2 ganglioside during ethanol-induced apoptotic neurodegeneration in the developing mouse brain. J Neurochem 121(4):649–661
15. Matyash V, Liebisch G, Kurzchalia TV, Shevchenko A, Schwudke D (2008) Lipid extraction by methyl-tert-butyl ether for high-throughput lipidomics. J Lipid Res 49(5): 1137–1146
16. Ellman GL, Courtney KD, Andres V Jr, Feather-Stone RM (1961) A new and rapid colorimetric determination of acetylcholinesterase activity. Biochem Pharmacol 7:88–95

Chapter 11

Isolation of Exosomes and Microvesicles from Cell Culture Systems to Study Prion Transmission

Pascal Leblanc, Zaira E. Arellano-Anaya, Emilien Bernard, Laure Gallay, Monique Provansal, Sylvain Lehmann, Laurent Schaeffer, Graça Raposo, and Didier Vilette

Abstract

Extracellular vesicles (EVs) are composed of microvesicles and exosomes. Exosomes are small membrane vesicles (40–120 nm sized) of endosomal origin released in the extracellular medium from cells when multivesicular bodies fuse with the plasma membrane, whereas microvesicles (i.e., shedding vesicles, 100 nm to 1 μm sized) bud from the plasma membrane. Exosomes and microvesicles carry functional proteins and nucleic acids (especially mRNAs and microRNAs) that can be transferred to surrounding cells and tissues and can impact multiple dimensions of the cellular life. Most of the cells, if not all, from neuronal to immune cells, release exosomes and microvesicles in the extracellular medium, and all biological fluids including blood (serum/plasma), urine, cerebrospinal fluid, and saliva contain EVs.

Prion-infected cultured cells are known to secrete infectivity into their environment. We characterized this cell-free form of prions and showed that infectivity was associated with exosomes. Since exosomes are produced by a variety of cells, including cells that actively accumulate prions, they could be a vehicle for infectivity in body fluids and could participate to the dissemination of prions in the organism. In addition, such infectious exosomes also represent a natural, simple, biological material to get key information on the abnormal PrP forms associated with infectivity.

In this chapter, we describe first a method that allows exosomes and microvesicles isolation from prion-infected cell cultures and in a second time the strategies to characterize the prions containing exosomes and their ability to disseminate the prion agent.

Key words Extracellular vesicles, Exosomes, Microvesicles, Prions, PrP^{C}, PrP^{Sc}, PrP^{Res}, Spreading

1 Introduction

Communication between cells and tissues can be mediated through different pathways including direct cell-cell contacts or through the release in the extracellular milieu, of functional signals composed of soluble mediators like hormones, neurotransmitters, or cytokines. However, the extracellular milieu was found to be more complex than previously thought. Indeed, characterization

Andrew F. Hill (ed.), *Exosomes and Microvesicles: Methods and Protocols,* Methods in Molecular Biology, vol. 1545, DOI 10.1007/978-1-4939-6728-5_11,

of biological fluids revealed the presence of additional components or cell-derived structures including nucleoprotein complexes (for reviews [1, 2]) and membrane vesicles (i.e., extracellular vesicles; EVs) released by the cells and tissues (for review [3]). Recent studies revealed that release of nucleoprotein complexes as well as extracellular vesicles plays a major role in the intercellular communication [1, 2, 4–6]. Here we will focus our attention on the recently identified extracellular vesicles, as powerful mediators of intercellular communication but also as potential structures involved in the transport and the dissemination of prion agents.

Most of cells (if not all) release EVs in the extracellular environment. EVs are composed of microvesicles (including shedding vesicles and apoptotic bodies) and exosomes (for reviews [3, 7]). Biogenesis of microvesicles occurs at the cell surface by a budding process of the plasma membrane similar to what is depicted for retroviruses. Despite this, the mechanisms involved in the budding process of shedding vesicles are still elusive although it was described that Arf6, actin [8], and more recently Tsg101 [9] could be involved. Exosome biogenesis, meanwhile, is initiated by a budding process of the limiting membrane surrounding the early endosomes and is ended with the formation of multivesicular bodies (MVBs, late endosomal compartments) when the lumen is filled with small intraluminal vesicles (ILVs) [10]. After fusion of the MVB surrounding membrane with the plasma membrane, the released ILVs in the extracellular medium are subsequently termed exosomes. While exosomes share size distribution with viruses (from 30 to 120 nm), microvesicles (especially the shedding vesicles) overlap in size with bacteria and protein aggregates (from 100 nm to 1 μm). Apoptotic bodies fall into the size range of 1–5 μm (for review [3]).

The cellular and molecular mechanisms by which exosomes are formed are a recent and very active field of investigation. Today, it is largely admitted that different subpopulations of MVBs/ILVs coexist in the cell cytosol, explaining, at least in part, the strong heterogeneity of size and composition of exosomes released in the extracellular medium [11–13]. Currently, three potential mechanisms have been identified to be involved in exosome biogenesis based on knowledge acquired for the ILVs formation and cargo sorting. The first one, well described in yeast, involves the endosomal sorting complex required for transport (ESCRT) machinery [14–16]. The second pathway involves the synthesis of ceramide [17], and the last process (ESCRT and ceramide independent) involves, in some cellular contexts, tetraspanin proteins such as CD63 or CD9 [11, 18, 19].

Microvesicles and exosomes have been isolated in most of biological fluids including cerebrospinal fluid [20, 21]; milk [22, 23]; saliva [24, 25]; urine [26]; semen [27]; nasal, ascites, bronchoalveolar, and amniotic fluids [28–31]; bile [32]; and blood (serum/plasma) [33] and are recovered in most of cell culture supernatants. Extracellular vesicles, especially exosomes, were discovered 30 years ago [34, 35] and were considered until recently little more than garbage cans whose job was to discard unwanted cellular components [36]. Over the past few years, evidence has revealed that these structures also act as messengers, conveying information to distant tissues [37]. Biochemical characterization of isolated microvesicles and exosomes revealed they can passively and actively recruit proteins and nucleic acids (especially mRNAs and microRNAs) that are functional and that can be transported and transferred to other cells and tissues, where they can alter cellular functions and physiology [38–43].

Transmissible spongiform encephalopathies (TSEs), or prion diseases, are a group of fatal neurodegenerative diseases affecting humans and animals. TSEs are characterized by the conformational conversion of the normally soluble, protease-sensitive cellular prion protein (PrP^{C}) into an abnormal detergent-insoluble and partially protease-resistant isoform called PrP^{Sc} [44]. PrP^{Sc} has been found to be tightly associated with prion infectivity, and recent evidence has shown that PrP^{Sc} is a major component of the prion infectious agent (for review [45]).

In TSEs, infectious prions generally enter the host through the gastrointestinal tract and are first detected in the periphery long before they reach the central nervous system where they induce the pathology. Different cell types, including lymphoid organs and peripheral nerves, contribute to the replication and transfer of infectious prions from peripheral sites to the brain, yet the molecular mechanisms underlying the cell-to-cell spreading of prions in vivo are unclear [46] although some hypotheses have been suggested [47–49].

In vitro experiments revealed that transfer of prions from infected cells to uninfected cells could be mediated through direct cell-cell contacts [50, 51] or tunneling nanotubes (TNTs) [52]. However, in 1997 Schatzl and colleagues also observed that prion infectivity could be detected in the conditioned cell culture medium [53] suggesting that prions can also be secreted. Interestingly, in precursor studies, we found that prion proteins (PrP^{C} and $PrPS^{c}$) can be released in the extracellular milieu in association with exosomes [54–57] but also with viral particles [56]. In addition, our study revealed that purified exosomes from supernatants of infected cell cultures can transmit the prion agent and induce the pathology in inoculated mice [55]. Association of PrP^{Sc} and PrP^{c} with exosomes was also confirmed in different studies [21, 58–67]. Recently, we and others observed an efficient inhibition of infectious prion multiplication and release by targeting the exosomal pathway [68–70]. These data suggest that multivesicular body compartments are internal sites of prion conversion and that the exosomal pathway could correspond to a potential therapeutical target to limit the spreading of prions.

In one study, prions were also found to be released in association with microvesicles [71] raising the possibility that prions can be released in the extracellular medium through several pathways.

While in vitro models indicate that transmission of prions from cell-to-cell can occur through secretion of infectious EV/exosomes, there is no evidence yet that this is also occurring in the in vivo situation.

Detection of infectious prions has been validated in several biological fluids including urine, saliva, milk, and blood of infected animals through different experimental strategies like in vivo bioassays, protein misfolding cyclic amplification (PMCA), or by real-time quaking-induced conversion (RT-QuIC; [72–88]). In this respect, recent studies revealed the association of pathological PrP with exosomes released in blood of infected animals [89, 90].

While TSEs have been considered for a long time to be the only neurodegenerative disorder involving an infectious agent of protein origin able to spread in the organism and between individuals, the last 10 years have seen the emergence of key proteins involved in other neurodegenerative disorders such as Alzheimer's disease, synucleinopathies (Lewy body dementia (LBD), Parkinson's disease, multisystemic atrophy (MSA)), or amyotrophic lateral sclerosis (ALS) that share some features with prions (for reviews [91–93]). β-amyloid peptide (Aβ), Tau, α-synuclein, super oxide dismutase 1 (SOD1), or Tar DNA-binding protein 43 (TDP-43) have been all found to be misfolded and aggregated in their respective disorders through seeded polymerization processes. Similarly to prions, these proteins in their pathological state are able to self-propagate, possibly leading to their spreading in the organism. The strong similarities between the misfolded proteins presented above and the prion agent involved in TSEs make it tempting to consider that SOD1/Tau/Aβ/TDP-43 or α-synuclein could correspond to prions [91, 92, 94–100]. However, TSE's prion agents are infectious and are transmissible between individuals, whereas other prion-like proteins (also called prionoids) did not so far presented the capacity to propagate within communities and cause macroepidemics such as kuru or bovine spongiform encephalopathy (for review [93]). More recently, the group of Stanley Prusiner identified the α-synuclein-MSA disease as a second potential prion disorder due to the accumulation and replication ex vivo and in vivo of a specific MSA strain of the α-synuclein protein [100, 101].

Interestingly, recent studies also revealed that, like prions, these prions and prionoids can be found associated with extracellular vesicles (especially with exosomes) released in the extracellular medium, in cell culture systems, or in biological fluids like CSF or blood [6, 98, 102–109] thus strongly suggesting that exosomes could potentially be involved in their in vivo spreading.

The role of the exosomal and/or the microvesicle pathways in the spreading of prions through the organism need to be deeply investigated. This approach involves the study of exosomes and microvesicles in prion-infected cell culture systems to understand and to characterize their roles in prion disease and more generally in neurodegenerative disorders associated with protein misfolding.

2 Materials

2.1 Reagents

Antibodies:

1. Prion protein antibodies (J. Grassi; SPI-Bio Bertin Pharma).
   (a) SAF32 (epitope 79–92; 1/10,000 for Western blotting (WB); 1/200 in IEM).
   (b) SAF83 (epitope 142–160; 1/200 in IEM).
   (c) Sha31 (epitope 145–152; 1/6000 for WB).
2. Tsg101 (ref. ARP37310 T100, Aviva System Biology; 1/1000 + 1% milk for WB).
3. Alix (Ref. pab0204, Covalab: 1/1000 for WB).
4. EF1α (Ref. 05-235 CBP-KK1, Upstate Cell Signaling; 1/500 for WB).
5. Calnexin (Ref. SPA-865, Stressgen; 1/1000 for WB).
6. Flotilin 1 (Ref. F1180, Sigma-Aldrich; 1/1000 for WB).
7. CD63 (ref H5C6 556019, BD Pharmingen; 1/300 + 0.5% milk for WB).
8. CD81 (ref JS-81 555675, BD Pharmingen; 1/500 + 0.5% milk for WB).
9. Bridging rabbit anti-mouse (Dakopatt) for immunogold electron microscopy (generally used at 1/200).
10. Protein-A-gold can be purchased from Cell Microscopy Center, Department of Cell Biology, Utrecht University, the Netherlands.

Cell culture:

1. Fetal calf serum (FCS) (cat. No. A15-102, GE Healthcare).
2. Opti-MEM Glutamax I (Ref. 51985-026, Gibco Life Technologies) containing 10% (w/v) FSC; 1% penicillin/streptomycin (Ref. 15140-122, Gibco Life Technologies 100 mg/ml).
3. Phosphate-buffered saline w/o calcium and magnesium (1× PBS) (Ref. SH30028.02, Hyclone).
4. Trypsin-EDTA (Ref. 15400-054, Gibco Life Technologies).
5. Doxycycline (ref. D1822, Sigma-Aldrich).

Biochemical reagents:

1. Proteinase K (PK) solution 20 mg/ml (cat. No.25530-049, Invitrogen).
2. PK-lysis buffer: 0.5 % sodium deoxycholate (Sigma-Aldrich, St Louis), 0.5 % Triton X-100, 150 mM NaCl, and 50 mM Tris–HCl, pH 7.5.
3. Phenylmethylsulfonyl fluoride (PMSF) (Sigma-Aldrich, St. Louis, MO) solubilized in 2-propanol (at 100 mM) and stored at −20 °C.
4. Sample buffer (reduced condition): 125 mM Tris–HCl, pH 6.8, 4 % sodium dodecyl sulfate (SDS), 10 % glycerol, 0.02 % bromophenol blue, and 5 % β-mercaptoethanol (Sigma-Aldrich, St. Louis).
5. Sample buffer (nonreduced condition): 125 mM Tris–HCl, pH 6.8, 4 % sodium dodecyl sulfate (SDS), 10 % glycerol, 0.02 % bromophenol blue.
6. Methanol.
7. Glycine (Sigma-Aldrich).
8. Uranyl acetate, pH 4.
9. Methyl cellulose-UA, pH 4: nine parts 2 % methyl cellulose and one part 4 % uranyl acetate (mix just before use).
10. Glutaraldehyde 25 % (EM grade; Electron Microscopy Sciences).
11. Paraformaldehyde 20 % (EM grade; Electron Microscopy Sciences).
12. Coomassie blue stain solution: 10 % acetic acid, 45 % methanol, and 0.025 % Coomassie blue R (Sigma-Aldrich).
13. BioRad protein assay (cat 500-0006).
14. Membrane for protein transfer Immobilon-P PVDF (cat. No. IPVH00010, Millipore).
15. Towbin transfer buffer: 25 mM Tris, 192 mM glycine (20 % methanol), pH 8.3.
16. MycoAlert kit (cat# LT07-118, Lonza).
17. TNE 10× buffer: 250 mM Tris–HCl pH 7.5; 1.5 M NaCl; 10 mM EDTA.

2.2 Equipment

1. Biosafety level 2 cell culture facility.
2. Tissue culture plasticware (corning, ref); 162 cm^2; T75 cm^2 vented flasks.
3. Cell culture incubator (37 °C, 5 % CO_2 atmosphere).
4. Centrifuge Beckman Allegra™ 25R/rotor (fixed angle) TA 10.250 (conical tubes 50 ml polypropylene).

5. Ultracentrifuge Beckman OPTIMA™ LE-80 K/swinging rotor Beckman SW32ti (tubes Beckman Ultra-Clear N°344058 or polyallomer N°326823); swinging rotor Beckman SW41ti (Ultra-Clear N°344059); fixed rotor 45Ti Beckman (tubes Herolab PC thick-walled threaded 94 ml ref 253290 with screw caps ref 254682).
6. Ultracentrifuge Beckman OPTIMA™ max/swinging rotor MLS-50 (tubes Beckman polyallomer N°326819).
7. Centrifuge (Jouan, CR412)/rotor M4 (cat 11174218).
8. Steritop Express TMPLUS filter 0.22 μm (Millipore).
9. Transmission electron microscope (CM120 electron microscope Philips).
10. Formvar-carbon coated EM grids [110].
11. Forceps (Dumont N°5).
12. Stainless steel loops (Pasteur "oese") [110]. Loops are slightly larger than the grids and are generally mounted on a 1 ml plastic micropipettor tip.
13. Whatman filter paper N°1.
14. Electrophoretic apparatus ATTO (AE-6450 dual mini Slab kit).
15. Trans-Blot semidry transfer apparatus (BioRad).

2.3 Cell Lines

1. N2a#58 and N2a#22L.
2. GT1-7#CT and scGT1-7#22L.
3. MovCT and scMov127S.
4. RovCT and scRov127S.
5. NIH3T3#CT and scNIH3T3#22L.

3 Methods

3.1 Cell Culture

Today, less than 30 cell lines and primary cells have been found to be permissive for prion replication [111]. Among them, we and others found that neuroblastoma cells [112–114], murine NIH3T3 fibroblasts [115], murine GT1-7 hypothalamic cell line [53, 114, 116], murine MovS6 Schwann-like cells [117], and the RK13 rabbit epithelial cells (ovRK13/Rov; [118]) when infected by prion agents release infectious prions in the extracellular medium in association with exosomes [54–59, 64, 65, 69]. N2a cells were also found to release infectious prions in association with bigger microvesicles [71].

MovS6 cells are neuroglial cells originating from dorsal root ganglia of transgenic mice expressing the ovine PrP^C (tgOv) and simian virus 40 (SV40) T-antigen. These cells were found to be permissive to a sheep scrapie agent (strain 127S). MovS6 cells were shown to sustain an efficient and stable replication of sheep prions based on the high level of accumulation of abnormal PrP (PrP^{Res}) and infectivity in infected cultures. Most of protocols described here have been carried out on the Mov cell line.

1. Culture of (murine and ovine) prion-infected cells is carried out in biosafety level 2 cell culture facility.
2. Most of the permissive cells described above are cultured in Opti-MEM medium containing 10% FCS, glutamine, and penicillin/streptomycin mix antibiotics (full Opti-MEM). The NIH3T3 cells are cultured in similar conditions excepted for the Opti-MEM that is replaced by DMEM. Permissive and prion-infected cells are easily available from laboratories working in the prion field. However, infection of permissive cells can also be established by treating cells with brain homogenates from prion-infected animals, lysates, or conditioned media from infected cell cultures (described in [119]). Once obtained, these cells can maintain prion replication over many passages.
3. Passaging of infected MovS6 cells is carried out by first washing the cells with PBS and by adding trypsin-EDTA for 2 min at room temperature. Cells are resuspended in full Opti-MEM medium and plated in adequate cell culture flasks/dishes. For routine culturing, normal and 127S-infected MovS6 cells are plated at 1:10 (*see* **Note 1**) and passaged twice per week (*see* **Note 2**).

3.2 Exosomes and Microvesicles Production

1. Before starting exosomes and microvesicles production, verify that your cells are free of mycoplasma using the MycoAlert kit (Lonza) (*see* **Note 3**).
2. Prepare ExoFree medium (*see* **Note 4**).
3. Plate out cells (2×10^6) in T162 cm^2 flasks in ExoFree medium (20 ml) (*see* **Note 5**).
4. Incubate cells for 4 days. Cells must not be overgrown at day 4 to avoid cell mortality or cell release in the cell conditioned medium (*see* **Note 6**).

3.3 Isolation of Exosomes and Microvesicles from Prion-Infected Cell Cultures

1. After 4 days of culture, conditioned media are collected (*see* **Note 5**), transferred into 50 ml Falcon tubes, and centrifuged at $300 \times g$ (swinging bucket M4, CR412 Jouan) for 5 min at 4 °C to gently remove cells in suspension.
2. Carefully collect the supernatant without pipetting the potential cell pellet and transfer supernatants into new 50 ml Falcon tubes and centrifuge at $2000 \times g$ (swinging bucket) for 20 min at 4 °C to remove cellular debris.
3. Carefully transfer the supernatant to new 50 ml tubes and centrifuge at $10{,}000 \times g$ (fixed angle rotor Allegra 25,

Beckman Coulter) for 30 min at 4 °C. At this step, visible pellets correspond to microvesicles (*see* **Note 7**).

4. To recover exosomes (*see* **Note 8**), the resulting supernatants are carefully transferred into ultracentrifuge tubes (Ultra-Clear or polyallomer tubes, Beckman Coulter) into a sterile environment (*see* **Note 9**) and ultracentrifuged at 100,000 × *g* ($g_{average}$) for 90 min at 4 °C in SW32ti swinging rotor in an OPTIMA™ LE-80 K ultracentrifuge (Beckman Coulter). Alternatively, if larger volumes are collected, you can use the 45Ti rotor (Beckman Coulter) (*see* **Note 10**).
5. Immediately after spin completion, discard the supernatant by inversion of the tube. Note that at this step, a pellet, enriched in exosomes, can be visible in the center of the bottom of the ultracentrifuge tube.
6. Resuspend the individual exosomal pellets in PBS by up and down pipetting (*see* **Note 11**).
7. Pool the resuspended exosomes and transfer into a new ultracentrifuge tube and centrifuge at 100,000 × *g* for 75 min at 4 °C. This step can be realized in an SW41Ti rotor or small swinging rotors like SW60Ti or MLS-50 (Beckman Coulter).
8. Immediately after centrifugation, discard the supernatant by inversion. At this step a translucent pellet can be visible in the center of the bottom of the tube.
9. Using an aspirating pipet covered with a sterile tip, aspirate the small remaining droplets of PBS on the walls of the tubes. Do not suck the wall of the lower part of the tube to avoid the exosomal pellet loss.
10. Resuspend the pellet in specific resuspension buffers (*see* **Note 12**). This exosomal pellet will be named hereafter 100K pellet.
11. Use immediately this 100K pellet or store at −20 °C or at −80 °C for longer storage.

3.4 Characterization of Infectious Exosomes Using Linear Sucrose Density Gradients

Sucrose density gradients are usually used to isolate the exosomes. Exosomes fractionated through sucrose gradients are distributed in fractions ranging from 1.10 to 1.20 g/ml depending of the cell type from which they were isolated. The methodology we used is based on what has been described for retroviruses that display similar densities.

1. Prepare the linear sucrose gradient using two sucrose solutions corresponding to a 60% (weight/weight) and a 5% (w/w) sucrose in 1× TNE buffer.
2. Place 6 ml of 60% sucrose solution into the bottom of an Ultra-Clear ultracentrifuge tube (SW41Ti rotor, Beckman Coulter).

3. Overlay carefully >6 ml of the 5% sucrose solution onto the dense 60% sucrose phase and obtain a convex meniscus on the top of the two step gradient.
4. Close the tube with Parafilm, and maintain the Parafilm using a ring corresponding to a cut part of the top of a 15 ml Falcon tube. Place the same ring to the bottom of the gradient tube to equilibrate the tube. No air bubbles must be present inside the tube (very important; *see* **Note 13**).
5. Gently lay down the tube on its side at the horizontal and allow to diffuse 3 h at room temperature or overnight at 4 °C. Tubes must be maintained stably in this position.
6. Slowly straighten the tube to the vertical. The gradient is ready to use (*see* **Note 14**).
7. Take off the Parafilm.
8. Slowly pipet off 1.5 ml of sucrose gradient (*see* **Note 15**).
9. Gently overlay 0.5 ml of the exosomal suspension (*see* Subheading 3.3) on the preformed gradient.
10. Do not forget to make an equilibrium tube in the same conditions but with exosomal suspension from noninfected cells.
11. Ultracentrifuge the gradient at 100,000 × g ($g_{average}$) for 16–19 h at 4 °C to the equilibrium. Use medium acceleration and break to avoid disturbance of the gradient.
12. After spin completion, 15 fractions (about 740 μl each) are collected from the top of the tube using a pipet. Take 20 μl of each fraction and measure the refractive index to determine the density (*see* **Note 16**).
13. Take 80 μl of each fraction and solubilize the proteins with 20 μl of 5× sample buffer. Store the non-treated fractions at −20 °C or for longer storage at −80 °C.
14. Boil the samples at 95 °C for 5 min and briefly centrifuge the tubes.
15. Analyze the different fractions (30 μl of the samples) directly using SDS-polyacrylamide gel electrophoresis.
16. Transfer the gel onto a PVDF membrane (Towbin buffer containing 20% methanol) using a semidry Trans-Blot apparatus (BioRad).
17. Immunoblot the membrane with antibodies directed against exosomal markers like Alix, TSG101, cyclophilin A (CypA), elongation factor 1 alpha (EF1α), or flotillin and with PrP antibodies like SAF32 and Sha31.
18. If you want to use antibodies that recognize the exosomal markers CD63, CD81, or CD9 tetraspanin proteins, do not reduce and do not denature the samples before SDS-PAGE analysis (*see* **Note 17**).

19. If fractionated exosomes need to be analyzed in other buffer conditions, positive fractions for PrP^{C}/PrP^{Res} and exosomal markers can be pooled or not and diluted in 1× PBS and ultracentrifuged during 70 min at 100,000 × *g*. The exosomal pellets can thus be resuspended in the appropriate buffer.

3.5 Electron Microscopy

Purified extracellular vesicle preparations can be visualized by electron microscopy (EM). Under EM, exosomes appear with a "cup-shaped" structure and have sizes ranging from 50 to 120 nm. On the contrary, microvesicles display sizes of >120 nm with aspect of large membranous structures. Note that "cup-shaped" structures of exosomes correspond to artifactual structures due to fixation steps. Exosome pellets analyzed by cryoelectron microscopy display perfect circular structures [10].

1. Concentrated or purified exosomes (described in Subheadings 3.3 and 3.4, **step 1**, respectively) are fixed in PBS containing 2% PFA and 0.065% glutaraldehyde (EM grade) for 5 min at room temperature.
2. Fixed exosomal preparations (5–20 μl depending on the concentration of vesicles) are applied on Formvar-carbon-coated grids incubated for 30 min at room temperature.
3. Using clean forceps, transfer the grids on a drop of 1× PBS on a Parafilm, and wash the grids by floating them for 4 min, four times at room temperature.
4. Wash the grids by floating on a drop of bidistilled and filtered water eight times for 1 min on Parafilm.
5. During the washing steps, prepare the contrasting solution (4% uranyl acetate pH 4 + 2% methyl cellulose: 900 μl of methylcellulose + 100 μl of uranyl acetate pH 4).
6. Spot two drops of contrasting solution on a glass plate covered by a Parafilm and put the plate on ice.
7. Wash the grids in the first drop and then incubate in the dark on the second drop for 10 min.
8. Remove the excess of staining solution using a Pasteur "oese" (stainless steel loops) and strip (push the loop sideways) the "oese" + grids on a Whatman filter N°1.
9. Let the grids dry on the "oese" for 5–10 min at room temperature.
10. Take off the grids with forceps, and view the preparations with a transmission electron microscope at 80 kV (CM120 electron microscope Philips, Eindhoven, the Netherlands).

3.6 Detection of PrP^{Sc} and Infectivity in Extracellular Vesicles

Association of PrP^{Sc} with exosomes or microvesicles can be detected through different experimental strategies. The main method used to determine the presence of PrP^{Sc} is the detection of the partially Proteinase K (PK)-resistant PrP (PrP^{Res}) by Western blotting.

Alternatively, association of PrP^{Sc} with exosomes or microvesicles can also be detected by immunogold electron microscopy (IEM) in presence of guanidine isothiocyanate treatment [56, 120].

Prion infectivity is most of the time associated with the presence of PrP^{Res} , and our analyses showed that exosomes released by infected cells are infectious for animals. However, detection of PrP^{Res} is notoriously less sensitive than infectivity assays, and some infectious PrP^{Sc} entities are sensitive to PK digestion [121]. Therefore, it is highly recommended to confirm the presence of exosome-associated prions by performing infectivity assays. For decades, the gold standard for detection and quantification of prion infectivity had relied on inoculation of appropriate animals and determination of the time to terminal disease (incubation period method) or the dilution at which a given sample transmits disease to 50% of the inoculated animals (endpoint titration method). More recently, alternative strategies have been developed leading to simple, versatile, and sensitive cell-based assays (SCA) to detect prion infectivity [122, 123].

3.6.1 Immunoblot Detection of PrP^{Res}

For negative and positive controls, use noninfected and infected cells, respectively. Ideally, use the exosome producer cells. Typically, PrP^{Res} can be detected in p100K harvested from 5 ml of conditioned medium.

1. Lyse the cells and exosomes (p100K or purified exosomes from density gradients) in ice-cold PK-lysis buffer during 20 min.
2. Centrifuge the cell lysates at 800 × *g* for 10 min at 4 °C to pellet the nuclei and insoluble materials.
3. Collect the post-nuclear supernatant.
4. Determine the protein concentration using a BCA or Bradford assay.
5. Digest 300 μg of total proteins with PK (16 μg of PK per mg of protein) for 30 min at 37 °C in a final volume of 300 μl of PK-lysis buffer (*see* **Note 18**).
6. Do not forget to reserve cell and exosomal lysates for non-PK-treated controls.
7. After 30 min, incubate the digestions at 4 °C for 5 min.
8. Add PMSF to a final concentration of 5 mM (*see* **Note 19**) and incubate for 5 min on ice.
9. Centrifuge samples at 20,000 × *g* for 45 min at 4 °C (*see* **Note 20**).
10. Carefully discard the supernatant using a pipet (P200). Do not touch the white pellet with the tip (*see* **Note 21**).
11. Resuspend the pellets in 15 μl of 2× sample buffer, vortex, and boil for 5 min at 95 °C.

12. In the same time denature non-PK treated cell and exosomal lysates from (6) with sample buffer. Use these samples as non-PK-treated controls (*see* **Note 17**).
13. Analyze the samples by SDS-PAGE and immunoblotting.

3.6.2 PrP^{Sc} Immunogold Electron Microscopy (IEM) Detection

1. Concentrated or purified exosomes are fixed in 1× PBS containing 2% PFA and 0.065% glutaraldehyde (EM grade) and incubated for 5 min at room temperature.
2. Fixed exosomal preparations are applied on Formvar-carbon-coated grids (5–20 μl depending on the concentration of vesicles) for 30 min at room temperature.
3. Wash the grids by floating on a drop of 1× PBS containing 50 mM glycine for four times of 4 min each at room temperature on a sheet of Parafilm.
4. Treat the grids with 3 M guanidine thiocyanate for 5 min at room temperature for PrP^{Sc} epitope revelation (*see* **Note 8**).
5. Wash the grids eight times (for 1 min) in 1× PBS to eliminate the guanidine thiocyanate.
6. Incubate the grids on a drop of 1× PBS containing 10% FCS to saturate.
7. Incubate the grids with the anti-PrP SAF83 or Sha31 (1/200) antibodies in 1× PBS 5% FCS for 30 min at room temperature (*see* **Note 8**).
8. Wash six times for 3 min in 1× PBS containing 0.5% FCS.
9. Incubate the grids with a bridging rabbit anti-mouse IgM antibody (1/200) 30 min at room temperature in 1× PBS 5% FCS.
10. Wash six times for 3 min with 1× PBS containing 0.5% FCS at room temperature.
11. Incubate the grids with the protein-A-gold 15 nm (1/60) in 1× PBS containing 5% FCS for 20 min at room temperature.
12. Wash the grids six times for 3 min in 1× PBS containing 0.5% FCS at room temperature.
13. Fix the complex antigen-antibody-protein-A-gold with 1× PBS containing 1% glutaraldehyde for 5 min at room temperature.
14. Wash four times for 4 min in 1× PBS containing 50 mM glycine at room temperature.
15. Wash the grids by floating on a drop of bidistilled and filtered water eight times for 1 min on Parafilm.
16. During the washing steps, prepare the contrasting solution.
17. Spot two drops of contrasting solution on a glass plate covered by a Parafilm and put the plate on ice.

18. Wash the grids in the first drop and then incubate in the dark on the second drop for 10 min.
19. Remove the excess of staining solution using a Pasteur "oese" (stainless steel loops) and strip (push the loop sideways) the "oese" + grids on a Whatman filter N°1.
20. Let the grids to dry on the "oese" for 5–10 min at room temperature.
21. View the preparations with a transmission electron microscope (CM120 Electron Microscope; Philips, Eindhoven, the Netherlands).

3.6.3 Cell-Based Quantification of Prion Infectivity in Cellular and Exosome Fractions

Reliable and sensitive detection of 127S ovine prion infectivity can be conveniently obtained by exposing recipient permissive ovRK13 cells to samples to be tested [122].

Cell homogenates are infected Mov cell cultures used to prepare the conditioned media. Typically, cells from a single T-162 cm^2 flask are trypsinized and pelleted at 120 × *g* for 7 min. Cell pellets are resuspended in 1 ml of PBS and homogenized in a high-speed homogenizer. Alternatively, the cell suspensions are subjected to 4 cycles of −80 °C freezing/37 °C thawing. Protein concentration is determined by BCA. The exosome fractions are either exosome pellets resuspended in PBS or pellets from sucrose fractions resuspended into PBS. As a general guideline, infectivity in cell homogenate samples containing a few micrograms of proteins and in exosome pellets harvested from the equivalent of 1.5 ml of conditioned medium is typically detected by the ovRK13 cell-based assay.

1. Seed ovRK13 cells in six-well plates ($0.5–0.75 \times 10^6$ cells per well in 3 ml of medium containing 1 μg/ml of doxycycline (dox)), and incubate at 37 °C, 5% CO_2, until the cells are confluent (2 days).
2. Remove the cell culture medium and inoculate the cells by adding 3 ml of dox-supplemented medium containing different dilutions of the samples to be tested (cell homogenate, exosomes). We usually test 1/1, 1/3, 1/9, and 1/81 dilutions of each sample on target cells (*see* Fig. 1).
3. Incubate the cells for 7 days.
4. Aspirate the medium and add 3 ml of fresh dox-containing medium.
5. Change the medium in each inoculated well every 5–7 days.
6. Twenty-eight days after cell inoculation, remove the medium and rinse the monolayers with 3 ml of cold 1× PBS.
7. Add 1 ml of PK-lysis buffer to each well.
8. Incubate for 10 min at 4 °C.
9. Collect the whole lysates in test tubes and clarify by centrifugation at 425 × *g*, 1 min in a microcentrifuge at 4 °C.

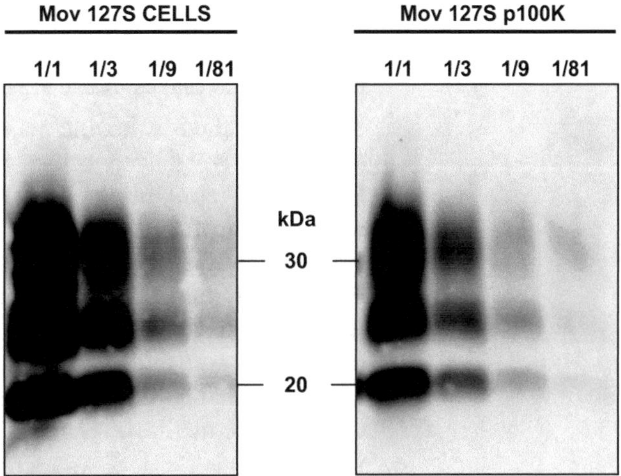

Fig. 1 Cell-based assay of infectivity in p100K and cell fractions. Serial dilutions (from 1 to 1/81) of Mov 127S cell homogenates and the corresponding p100K were inoculated to ovRK13 cells. Four weeks later, infection was visualized by immunodetection of PrP^{Res} in the inoculated cultures

10. Collect the post-nuclear supernatants in clean tubes.
11. Quantify the cellular proteins by BCA assay and store at −20 °C until processing for abnormal PrP detection by immunoblotting.
12. For PrP^{Res} detection, incubate 500–750 μg of cellular proteins with PK (1 μg of PK per 250 μg of proteins) for 2 h at 37 °C.
13. Stop the reaction with Pefabloc (4 mM final).
14. Centrifuge the samples at 20,000 × *g* for 30 min at room temperature in a microfuge.
15. Discard the supernatant.
16. Add 20–30 μl of sample buffer to the barely visible pellet and denature the sample at 100 °C for 10 min.
17. Store at −20 °C until PrP^{Res} analysis by immunoblotting.

In this assay, the amount of PrP^{Res} accumulating in the inoculated cells is proportional to the amount of 127S infectivity in the inoculum (*see* Fig. 1). By comparing PrP^{Res} levels in cells inoculated by appropriate dilutions, one can estimate the relative infectious titer of the cellular, exosome pellet and gradient fractions.

4 Notes

1. We observed that repeated passaging of infected MovS6 cells with higher dilutions leads to a significant loss of PrP^{Res}.
2. In our hands, passaging of infected MovS6 cells only one time per week at 1:10 dilution leads to high cell mortality and fusiform cell phenotype.

3. Mycoplasma contaminations can strongly affect the cell biology (such as cell division, viability) and thus the release of extracellular vesicles in the extracellular medium.
4. FCS is enriched in bovine exosomes which must be removed to selectively analyze cell-secreted exosomes. To this end, Opti-MEM containing 20% FCS (420 ml, e.g., 6×70 ml) must be centrifuged overnight (16 h) at 100,000×g ($g_{average}$) at 4 °C in the 45Ti rotor (Beckman Coulter). The resulting 20% FCS-depleted medium (referred to as ExoFree medium) is then filtered through a 0.22 μm Steritop Express™ PLUS (Millipore) and diluted to the half with Opti-MEM to reach 10% of FCS. This method allows preparing 840 ml of ready to use ExoFree medium. The antibiotic mix is added before use.
5. For classical studies, we use 2×T162 cm^2 flasks corresponding to a final volume of 40 ml of supernatant. For proteomic or transcriptomic analyses of normal and infected Mov p100K, we use 10×T162 cm^2 of noninfected and infected cells corresponding to 200 ml of supernatant for each condition.
6. Ideally, cells must be just at confluence, but not overgrown to avoid cell lethality and release of artifactual shedding vesicles and apoptotic bodies in the conditioned medium.
7. Incline the tube in the opposite and slowly pipet the supernatant.
8. At this step, many protocols use a 0.22 μm filtration to eliminate contaminations by vesicles with sizes >200 nm. We don't perform this step because it is well known that PrP^{Sc} is sticky and filtration is expected to result in PrP^{Sc} loss.
9. If exosomes containing PrP^{Sc} are used as inoculum to infect permissive target cells, the isolation of exosomes must be carried out in a sterile environment to avoid contamination by microorganisms.
10. When larger volumes (2×200 ml) of supernatants are handled (e.g., proteomic and transcriptomic studies of normal versus infected cells), we perform the 100,000×g ultracentrifugation in larger volume tubes (*see* Subheading 2) with the 45Ti rotor (Beckman Coulter). Using this rotor, you can ultracentrifuge easily a total of 400 ml of supernatant corresponding to 200 ml of conditioned medium from noninfected cells and 200 ml from infected cells. Note that although this rotor is useful to ultracentrifuge larger volumes, this rotor is an angle fixed and the resulting p100K pellet is thus less stable and less easy to recover.
11. We usually use sterile filter tips to avoid contamination of pipets with infectious materials.
12. Depending on the downstream experiments, you have to select the appropriate buffer for resuspension. For example, fractionation of exosomes through linear sucrose density or iodixanol

velocity gradients or cell-based assay of infectivity will need resuspension of the exosomal pellet in PBS. For detection of the Proteinase K (PK)-resistant PrP^{Res}, the p100K may be resuspended in PK-lysis buffer. Similarly, proteomic or transcriptomic analyses of the final exosomal pellet will require a specific buffer.

13. Bubbles inside the tube will destroy the linearity of the gradient during its formation.
14. Gradients preformed in these conditions are simple to make and highly reproducible.
15. Elimination of 1.5 ml of the top of the gradient allows overlying the exosomal suspension (0.5 ml) and letting free about 1 cm of the top of the tube. Once realized, we consistently obtain 8–60% linear sucrose gradients.
16. If you use prion-infected materials (exosomes from infected cell cultures), note that you have to deeply decontaminate your refractometer after use. On the other hand, you can measure the refractive index on a parallel linear density gradient that has been loaded with exosomal preparation from a normal cell culture. In our hands, this alternative gave good results.
17. For detection of CD63, CD81, and CD9 tetraspanin proteins as exosomal marker, do not use reducing agents (i.e., β-mercaptoethanol and/or DTT), and do not boil the samples.
18. Depending on the volume of conditioned medium used, the amount of proteins in the exosomal lysates may be very low. It is advisable to adjust the protein concentration to 300 μg with cell lysates from noninfected cells.
19. Addition of PMSF at 4 °C to block the PK digestion leads to the formation of white flakes that correspond to PMSF precipitation at this temperature.
20. White flakes make a white pellet after centrifugation. PrP^{Res} is present in this pellet.
21. PrP^{Res} is very sticky, so do not touch this pellet with the tip to avoid PrP^{Res} loss.
22. The SAF83 or the Sha31 monoclonal antibodies are directed against the C-terminal part of the PrP protein (epitopes 142–160 and145–152, respectively). They allow the detection of PrP^{Sc} after denaturation by the guanidine thiocyanate.

References

1. Chen X, Liang H, Zhang J, Zen K, Zhang CY (2012) Horizontal transfer of microRNAs: molecular mechanisms and clinical applications. Protein Cell 3(1):28–37. doi:10.1007/s13238-012-2003-z
2. Turchinovich A, Weiz L, Burwinkel B (2012) Extracellular miRNAs: the mystery of their origin and function. Trends Biochem Sci 37(11): 460–465, doi:S0968-0004(12)00115-6 [pii] 10.1016/j.tibs.2012.08.003

3. Gyorgy B, Szabo TG, Pasztoi M, Pal Z, Misjak P, Aradi B, Laszlo V, Pallinger E, Pap E, Kittel A, Nagy G, Falus A, Buzas EI (2011) Membrane vesicles, current state-of-the-art: emerging role of extracellular vesicles. Cell Mol Life Sci 68(16):2667–2688. doi:10.1007/s00018-011-0689-3
4. Bobrie A, Colombo M, Raposo G, Thery C (2011) Exosome secretion: molecular mechanisms and roles in immune responses. Traffic 12(12):1659–1668. doi:10.1111/j.1600-0854.2011.01225.x
5. Ludwig AK, Giebel B (2012) Exosomes: small vesicles participating in intercellular communication. Int J Biochem Cell Biol 44(1):11–15. doi:10.1016/j.biocel.2011.10.005, S1357-2725(11)00267-6 [pii]
6. Schneider A, Simons M (2013) Exosomes: vesicular carriers for intercellular communication in neurodegenerative disorders. Cell Tissue Res 352(1):33–47. doi:10.1007/s00441-012-1428-2
7. Cocucci E, Racchetti G, Meldolesi J (2009) Shedding microvesicles: artefacts no more. Trends Cell Biol 19(2):43–51. doi:10.1016/j.tcb.2008.11.003, S0962-8924(08)00283-3 [pii]
8. Muralidharan-Chari V, Clancy J, Plou C, Romao M, Chavrier P, Raposo G, D'Souza-Schorey C (2009) ARF6-regulated shedding of tumor cell-derived plasma membrane microvesicles. Curr Biol 19(22):1875–1885. doi:10.1016/j.cub.2009.09.059, S0960-9822(09)01772-2 [pii]
9. Nabhan JF, Hu R, Oh RS, Cohen SN, Lu Q (2012) Formation and release of arrestin domain-containing protein 1-mediated microvesicles (ARMMs) at plasma membrane by recruitment of TSG101 protein. Proc Natl Acad Sci U S A 109(11):4146–4151. doi:10.1073/pnas.1200448109, 1200448109 [pii]
10. Raposo G, Stoorvogel W (2013) Extracellular vesicles: exosomes, microvesicles, and friends. J Cell Biol 200(4):373–383. doi:10.1083/jcb.201211138, jcb.201211138 [pii]
11. Buschow SI, Nolte-'t Hoen EN, van Niel G, Pols MS, ten Broeke T, Lauwen M, Ossendorp F, Melief CJ, Raposo G, Wubbolts R, Wauben MH, Stoorvogel W (2009) MHC II in dendritic cells is targeted to lysosomes or T cell-induced exosomes via distinct multivesicular body pathways. Traffic 10(10):1528–1542. doi:10.1111/j.1600-0854.2009.00963.x, TRA963 [pii]
12. Mobius W, van Donselaar E, Ohno-Iwashita Y, Shimada Y, Heijnen HF, Slot JW, Geuze HJ (2003) Recycling compartments and the internal vesicles of multivesicular bodies harbor most of the cholesterol found in the endocytic pathway. Traffic 4(4):222–231, doi:072 [pii]
13. White IJ, Bailey LM, Aghakhani MR, Moss SE, Futter CE (2006) EGF stimulates annexin 1-dependent inward vesiculation in a multivesicular endosome subpopulation. EMBO J 25(1):1–12. doi:10.1038/sj.emboj.7600759, 7600759 [pii]
14. Babst M (2005) A protein's final ESCRT. Traffic 6(1):2–9. doi:10.1111/j.1600-0854.2004.00246.x, TRA246 [pii]
15. Colombo M, Moita C, van Niel G, Kowal J, Vigneron J, Benaroch P, Manel N, Moita LF, Thery C, Raposo G (2013) Analysis of ESCRT functions in exosome biogenesis, composition and secretion highlights the heterogeneity of extracellular vesicles. J Cell Sci 2013:39p. doi:10.1242/jcs.128868, jcs.128868 [pii]
16. Henne WM, Buchkovich NJ, Emr SD (2011) The ESCRT pathway. Dev Cell 21(1):77–91. doi:10.1016/j.devcel.2011.05.015, S1534-5807(11)00207-3 [pii]
17. Trajkovic K, Hsu C, Chiantia S, Rajendran L, Wenzel D, Wieland F, Schwille P, Brugger B, Simons M (2008) Ceramide triggers budding of exosome vesicles into multivesicular endosomes. Science 319(5867):1244–1247, doi:319/5867/1244 [pii] 10.1126/science.1153124
18. Perez-Hernandez D, Gutierrez-Vazquez C, Jorge I, Lopez-Martin S, Ursa A, Sanchez-Madrid F, Vazquez J, Yanez-Mo M (2013) The intracellular interactome of tetraspanin-enriched microdomains reveals their function as sorting machineries toward exosomes. J Biol Chem 288(17):11649–11661. doi:10.1074/jbc.M112.445304, M112.445304 [pii]
19. van Niel G, Charrin S, Simoes S, Romao M, Rochin L, Saftig P, Marks MS, Rubinstein E, Raposo G (2011) The tetraspanin CD63 regulates ESCRT-independent and -dependent endosomal sorting during melanogenesis. Dev Cell 21(4):708–721. doi:10.1016/j.devcel.2011.08.019, S1534-5807(11)00357-1 [pii]
20. Street JM, Barran PE, Mackay CL, Weidt S, Balmforth C, Walsh TS, Chalmers RT, Webb DJ, Dear JW (2012) Identification and proteomic profiling of exosomes in human cerebrospinal fluid. J Transl Med 10:5. doi:10.1186/1479-5876-10-5, 1479-5876-10-5 [pii]
21. Vella LJ, Greenwood DL, Cappai R, Scheerlinck JP, Hill AF (2008) Enrichment of prion protein in exosomes derived from ovine

cerebral spinal fluid. Vet Immunol Immunopathol 124(3-4):385–393. doi:10.1016/j.vetimm.2008.04.002, S0165-2427(08)00165-7 [pii]

22. Admyre C, Johansson SM, Qazi KR, Filen JJ, Lahesmaa R, Norman M, Neve EP, Scheynius A, Gabrielsson S (2007) Exosomes with immune modulatory features are present in human breast milk. J Immunol 179(3): 1969–1978
23. Lasser C, Alikhani VS, Ekstrom K, Eldh M, Paredes PT, Bossios A, Sjostrand M, Gabrielsson S, Lotvall J, Valadi H (2011) Human saliva, plasma and breast milk exosomes contain RNA: uptake by macrophages. J Transl Med 9:9. doi:10.1186/1479-5876-9-9, 1479-5876-9-9 [pii]
24. Michael A, Bajracharya SD, Yuen PS, Zhou H, Star RA, Illei GG, Alevizos I (2010) Exosomes from human saliva as a source of microRNA biomarkers. Oral Dis 16(1):34–38. doi:10.1111/j.1601-0825.2009.01604.x, ODI1604 [pii]
25. Palanisamy V, Sharma S, Deshpande A, Zhou H, Gimzewski J, Wong DT (2010) Nanostructural and transcriptomic analyses of human saliva derived exosomes. PLoS One 5(1), e8577. doi:10.1371/journal.pone.0008577
26. Pisitkun T, Shen RF, Knepper MA (2004) Identification and proteomic profiling of exosomes in human urine. Proc Natl Acad Sci U S A 101(36):13368–13373. doi:10.1073/pnas.0403453101, 0403453101 [pii]
27. Poliakov A, Spilman M, Dokland T, Amling CL, Mobley JA (2009) Structural heterogeneity and protein composition of exosome-like vesicles (prostasomes) in human semen. Prostate 69(2):159–167. doi:10.1002/pros.20860
28. Admyre C, Grunewald J, Thyberg J, Gripenback S, Tornling G, Eklund A, Scheynius A, Gabrielsson S (2003) Exosomes with major histocompatibility complex class II and co-stimulatory molecules are present in human BAL fluid. Eur Respir J 22(4):578–583
29. Andre F, Schartz NE, Movassagh M, Flament C, Pautier P, Morice P, Pomel C, Lhomme C, Escudier B, Le Chevalier T, Tursz T, Amigorena S, Raposo G, Angevin E, Zitvogel L (2002) Malignant effusions and immunogenic tumour-derived exosomes. Lancet 360(9329):295–305. doi:10.1016/S0140-6736(02)09552-1, S0140-6736(02)09552-1 [pii]
30. Asea A, Jean-Pierre C, Kaur P, Rao P, Linhares IM, Skupski D, Witkin SS (2008) Heat shock protein-containing exosomes in mid-trimester amniotic fluids. J Reprod Immunol 79(1): 12–17. doi:10.1016/j.jri.2008.06.001, S0165-0378(08)00069-7 [pii]
31. Lasser C, O'Neil SE, Ekerljung L, Ekstrom K, Sjostrand M, Lotvall J (2011) RNA-containing exosomes in human nasal secretions. Am J Rhinol Allergy 25(2):89–93. doi:10.2500/ajra.2011.25.3573, 3573 [pii]
32. Masyuk AI, Huang BQ, Ward CJ, Gradilone SA, Banales JM, Masyuk TV, Radtke B, Splinter PL, LaRusso NF (2010) Biliary exosomes influence cholangiocyte regulatory mechanisms and proliferation through interaction with primary cilia. Am J Physiol Gastrointest Liver Physiol 299(4):G990–G999. doi:10.1152/ajpgi.00093.2010, ajpgi.00093.2010 [pii]
33. Caby MP, Lankar D, Vincendeau-Scherrer C, Raposo G, Bonnerot C (2005) Exosomal-like vesicles are present in human blood plasma. Int Immunol 17(7):879–887. doi:10.1093/intimm/dxh267, dxh267 [pii]
34. Harding C, Heuser J, Stahl P (1984) Endocytosis and intracellular processing of transferrin and colloidal gold-transferrin in rat reticulocytes: demonstration of a pathway for receptor shedding. Eur J Cell Biol 35(2):256–263
35. Pan BT, Teng K, Wu C, Adam M, Johnstone RM (1985) Electron microscopic evidence for externalization of the transferrin receptor in vesicular form in sheep reticulocytes. J Cell Biol 101(3):942–948
36. Thery C (2011) Exosomes: secreted vesicles and intercellular communications. F100 Biol Rep 3:15. doi:10.3410/B3-15
37. Simons M, Raposo G (2009) Exosomes--vesicular carriers for intercellular communication. Curr Opin Cell Biol 21(4):575–581. doi:10.1016/j.ceb.2009.03.007, S0955-0674(09)00077-5 [pii]
38. Al-Nedawi K, Meehan B, Micallef J, Lhotak V, May L, Guha A, Rak J (2008) Intercellular transfer of the oncogenic receptor EGFRvIII by microvesicles derived from tumour cells. Nat Cell Biol 10(5):619–624. doi:10.1038/ncb1725, ncb1725 [pii]
39. Ratajczak J, Miekus K, Kucia M, Zhang J, Reca R, Dvorak P, Ratajczak MZ (2006) Embryonic stem cell-derived microvesicles reprogram hematopoietic progenitors: evidence for horizontal transfer of mRNA and protein delivery. Leukemia 20(5):847–856. doi:10.1038/sj.leu.2404132, 2404132 [pii]
40. Ratajczak J, Wysoczynski M, Hayek F, Janowska-Wieczorek A, Ratajczak MZ (2006) Membrane-derived microvesicles:

important and underappreciated mediators of cell-to-cell communication. Leukemia 20(9):1487–1495. doi:10.1038/sj.leu.2404296, 2404296 [pii]

41. Skog J, Wurdinger T, van Rijn S, Meijer DH, Gainche L, Sena-Esteves M, Curry WT Jr, Carter BS, Krichevsky AM, Breakefield XO (2008) Glioblastoma microvesicles transport RNA and proteins that promote tumour growth and provide diagnostic biomarkers. Nat Cell Biol 10(12):1470–1476. doi:10.1038/ncb1800, ncb1800 [pii]
42. Valadi H, Ekstrom K, Bossios A, Sjostrand M, Lee JJ, Lotvall JO (2007) Exosome-mediated transfer of mRNAs and microRNAs is a novel mechanism of genetic exchange between cells. Nat Cell Biol 9(6):654–659. doi:10.1038/ncb1596, ncb1596 [pii]
43. Ridder K, Keller S, Dams M, Rupp AK, Schlaudraff J, Del Turco D, Starmann J, Macas J, Karpova D, Devraj K, Depboylu C, Landfried B, Arnold B, Plate KH, Hoglinger G, Sultmann H, Altevogt P, Momma S (2014) Extracellular vesicle-mediated transfer of genetic information between the hematopoietic system and the brain in response to inflammation. PLoS Biol 12(6), e1001874. doi:10.1371/journal.pbio.1001874, PBIOLOGY-D-14-00669 [pii]
44. Prusiner SB (1998) Prions. Proc Natl Acad Sci U S A 95(23):13363–13383
45. Soto C (2011) Prion hypothesis: the end of the controversy? Trends Biochem Sci 36(3):151–158. doi:10.1016/j.tibs.2010.11.001,S0968-0004(10)00210-0 [pii]
46. Aguzzi A, Heikenwalder M (2006) Pathogenesis of prion diseases: current status and future outlook. Nat Rev Microbiol 4(10):765–775. doi:10.1038/nrmicro1492, nrmicro1492 [pii]
47. Castro-Seoane R, Hummerich H, Sweeting T, Tattum MH, Linehan JM, Fernandez de Marco M, Brandner S, Collinge J, Klohn PC (2012) Plasmacytoid dendritic cells sequester high prion titres at early stages of prion infection. PLoS Pathog 8(2), e1002538. doi:10.1371/journal.ppat.1002538, PPATHOGENS-D-11-00640 [pii]
48. Klohn PC, Castro-Seoane R, Collinge J (2013) Exosome release from infected dendritic cells: a clue for a fast spread of prions in the periphery? J Infect 67(5):359–368. doi:10.1016/j.jinf.2013.07.024, S0163-4453(13)00211-9 [pii]
49. Kujala P, Raymond CR, Romeijn M, Godsave SF, van Kasteren SI, Wille H, Prusiner SB, Mabbott NA, Peters PJ (2011) Prion uptake in the gut: identification of the first uptake and replication sites. PLoS Pathog 7(12), e1002449. doi:10.1371/journal.ppat.1002449, PPATHOGENS-D-11-01207 [pii]
50. Kanu N, Imokawa Y, Drechsel DN, Williamson RA, Birkett CR, Bostock CJ, Brockes JP (2002) Transfer of scrapie prion infectivity by cell contact in culture. Curr Biol 12(7):523–530, doi:S0960982202007224 [pii]
51. Paquet S, Langevin C, Chapuis J, Jackson GS, Laude H, Vilette D (2007) Efficient dissemination of prions through preferential transmission to nearby cells. J Gen Virol 88(Pt 2): 706–713. doi:10.1099/vir.0.82336-0
52. Gousset K, Schiff E, Langevin C, Marijanovic Z, Caputo A, Browman DT, Chenouard N, de Chaumont F, Martino A, Enninga J, Olivo-Marin JC, Mannel D, Zurzolo C (2009) Prions hijack tunnelling nanotubes for intercellular spread. Nat Cell Biol 11(3):328–336. doi:10.1038/ncb1841, ncb1841 [pii]
53. Schatzl HM, Laszlo L, Holtzman DM, Tatzelt J, DeArmond SJ, Weiner RI, Mobley WC, Prusiner SB (1997) A hypothalamic neuronal cell line persistently infected with scrapie prions exhibits apoptosis. J Virol 71(11):8821–8831
54. Alais S, Simoes S, Baas D, Lehmann S, Raposo G, Darlix JL, Leblanc P (2008) Mouse neuroblastoma cells release prion infectivity associated with exosomal vesicles. Biol Cell 100(10):603–615. doi:10.1042/BC20080025, BC20080025 [pii]
55. Fevrier B, Vilette D, Archer F, Loew D, Faigle W, Vidal M, Laude H, Raposo G (2004) Cells release prions in association with exosomes. Proc Natl Acad Sci U S A 101(26):9683–9688. doi:10.1073/pnas.0308413101, 0308413101 [pii]
56. Leblanc P, Alais S, Porto-Carreiro I, Lehmann S, Grassi J, Raposo G, Darlix JL (2006) Retrovirus infection strongly enhances scrapie infectivity release in cell culture. EMBO J 25(12):2674–2685. doi:10.1038/sj.emboj.7601162, 7601162 [pii]
57. Arellano-Anaya ZE, Huor A, Leblanc P, Lehmann S, Provansal M, Raposo G, Andreoletti O, Vilette D (2015) Prion strains are differentially released through the exosomal pathway. Cell Mol Life Sci 72(6):1185–1196. doi:10.1007/s00018-014-1735-8
58. Bellingham SA, Coleman BM, Hill AF (2012) Small RNA deep sequencing reveals a distinct miRNA signature released in exosomes from prion-infected neuronal cells. Nucleic Acids Res 40(21):10937–10949. doi:10.1093/nar/gks832, gks832 [pii]

59. Coleman BM, Hanssen E, Lawson VA, Hill AF (2012) Prion-infected cells regulate the release of exosomes with distinct ultrastructural features. FASEB J 26(10):4160–4173. doi:10.1096/fj.11-202077, fj.11-202077 [pii]
60. Conde-Vancells J, Rodriguez-Suarez E, Gonzalez E, Berisa A, Gil D, Embade N, Valle M, Luka Z, Elortza F, Wagner C, Lu SC, Mato JM, Falcon-Perez M (2010) Candidate biomarkers in exosome-like vesicles purified from rat and mouse urine samples. Proteomics Clin Appl 4(4):416–425. doi:10.1002/prca.200900103
61. Faure J, Lachenal G, Court M, Hirrlinger J, Chatellard-Causse C, Blot B, Grange J, Schoehn G, Goldberg Y, Boyer V, Kirchhoff F, Raposo G, Garin J, Sadoul R (2006) Exosomes are released by cultured cortical neurones. Mol Cell Neurosci 31(4):642–648. doi:10.1016/j.mcn.2005.12.003, S1044-7431(05)00302-7 [pii]
62. Ritchie AJ, Crawford DM, Ferguson DJ, Burthem J, Roberts DJ (2013) Normal prion protein is expressed on exosomes isolated from human plasma. Br J Haematol 163(5):678–680. doi:10.1111/bjh.12543
63. Robertson C, Booth SA, Beniac DR, Coulthart MB, Booth TF, McNicol A (2006) Cellular prion protein is released on exosomes from activated platelets. Blood 107(10):3907–3911. doi:10.1182/blood-2005-02-0802, 2005-02-0802 [pii]
64. Vella LJ, Sharples RA, Lawson VA, Masters CL, Cappai R, Hill AF (2007) Packaging of prions into exosomes is associated with a novel pathway of PrP processing. J Pathol 211(5):582–590. doi:10.1002/path.2145
65. Vella LJ, Sharples RA, Nisbet RM, Cappai R, Hill AF (2008) The role of exosomes in the processing of proteins associated with neurodegenerative diseases. Eur Biophys J 37(3):323–332. doi:10.1007/s00249-007-0246-z
66. Wang G, Zhou X, Bai Y, Zhang Z, Zhao D (2010) Cellular prion protein released on exosomes from macrophages binds to Hsp70. Acta Biochim Biophys Sin (Shanghai) 42(5):345–350
67. Wang GH, Zhou XM, Bai Y, Yin XM, Yang LF, Zhao D (2011) Hsp70 binds to PrPC in the process of PrPC release via exosomes from THP-1 monocytes. Cell Biol Int 35(6):553–558. doi:10.1042/CBI20090391, CBI20090391 [pii]
68. Guo BB, Bellingham SA, Hill AF (2015) The neutral sphingomyelinase pathway regulates packaging of the prion protein into exosomes. J Biol Chem 290(6):3455–3467. doi:10.1074/jbc.M114.605253, M114.605253 [pii]
69. Vilette D, Laulagnier K, Huor A, Alais S, Simoes S, Maryse R, Provansal M, Lehmann S, Andreoletti O, Schaeffer L, Raposo G, Leblanc P (2015) Efficient inhibition of infectious prions multiplication and release by targeting the exosomal pathway. Cell Mol Life Sci 72(22):4409–4427. doi:10.1007/s00018-015-1945-8, 10.1007/s00018-015-1945-8
70. Yim YI, Park BC, Yadavalli R, Zhao X, Eisenberg E, Greene LE (2015) The multivesicular body is the major internal site of prion conversion. J Cell Sci 128(7):1434–1443. doi:10.1242/jcs.165472, jcs.165472 [pii]
71. Mattei V, Barenco MG, Tasciotti V, Garofalo T, Longo A, Boller K, Lower J, Misasi R, Montrasio F, Sorice M (2009) Paracrine diffusion of PrP(C) and propagation of prion infectivity by plasma membrane-derived microvesicles. PLoS One 4(4), e5057. doi:10.1371/journal.pone.0005057
72. Chen B, Morales R, Barria MA, Soto C (2010) Estimating prion concentration in fluids and tissues by quantitative PMCA. Nat Methods 7(7):519–520. doi:10.1038/nmeth.1465, nmeth.1465 [pii]
73. Henderson DM, Davenport KA, Haley NJ, Denkers ND, Mathiason CK, Hoover EA (2015) Quantitative assessment of prion infectivity in tissues and body fluids by real-time quaking-induced conversion. J Gen Virol 96(Pt 1):210–219. doi:10.1099/vir.0.069906-0, vir.0.069906-0 [pii]
74. Lacroux C, Simon S, Benestad SL, Maillet S, Mathey J, Lugan S, Corbiere F, Cassard H, Costes P, Bergonier D, Weisbecker JL, Moldal T, Simmons H, Lantier F, Feraudet-Tarisse C, Morel N, Schelcher F, Grassi J, Andreoletti O (2008) Prions in milk from ewes incubating natural scrapie. PLoS Pathog 4(12), e1000238. doi:10.1371/journal.ppat.1000238
75. Mathiason CK, Powers JG, Dahmes SJ, Osborn DA, Miller KV, Warren RJ, Mason GL, Hays SA, Hayes-Klug J, Seelig DM, Wild MA, Wolfe LL, Spraker TR, Miller MW, Sigurdson CJ, Telling GC, Hoover EA (2006) Infectious prions in the saliva and blood of deer with chronic wasting disease. Science 314(5796):133–136, doi:314/5796/133 [pii] 10.1126/science.1132661
76. Moda F, Gambetti P, Notari S, Concha-Marambio L, Catania M, Park KW, Maderna E, Suardi S, Haik S, Brandel JP, Ironside J, Knight R, Tagliavini F, Soto C (2014) Prions in the urine of patients with variant Creutzfeldt-Jakob disease. N Engl J Med 371(6):530–539. doi:10.1056/NEJMoa1404401

77. Murayama Y, Masujin K, Imamura M, Ono F, Shibata H, Tobiume M, Yamamura T, Shimozaki N, Terao K, Yamakawa Y, Sata T (2014) Ultrasensitive detection of PrP(Sc) in the cerebrospinal fluid and blood of macaques infected with bovine spongiform encephalopathy prion. J Gen Virol 95(Pt 11):2576–2588. doi:10.1099/vir.0.066225-0, vir.0.066225-0 [pii]
78. Notari S, Qing L, Pocchiari M, Dagdanova A, Hatcher K, Dogterom A, Groisman JF, Lumholtz IB, Puopolo M, Lasmezas C, Chen SG, Kong Q, Gambetti P (2012) Assessing prion infectivity of human urine in sporadic Creutzfeldt-Jakob disease. Emerg Infect Dis 18(1):21–28. doi:10.3201/eid1801.110589
79. Orru CD, Groveman BR, Hughson AG, Zanusso G, Coulthart MB, Caughey B (2015) Rapid and sensitive RT-QuIC detection of human Creutzfeldt-Jakob disease using cerebrospinal fluid. MBio 6(1):pii e02451-14. doi:10.1128/mBio.02451-14
80. Andreoletti O, Litaise C, Simmons H, Corbiere F, Lugan S, Costes P, Schelcher F, Vilette D, Grassi J, Lacroux C (2012) Highly efficient prion transmission by blood transfusion. PLoS Pathog 8(6), e1002782. doi:10.1371/journal.ppat.1002782, PPATHOGENS-D-11-02296 [pii]
81. Brown P, Rohwer RG, Dunstan BC, MacAuley C, Gajdusek DC, Drohan WN (1998) The distribution of infectivity in blood components and plasma derivatives in experimental models of transmissible spongiform encephalopathy. Transfusion 38(9):810–816
82. Gregori L, Kovacs GG, Alexeeva I, Budka H, Rohwer RG (2008) Excretion of transmissible spongiform encephalopathy infectivity in urine. Emerg Infect Dis 14(9):1406–1412. doi:10.3201/eid1409.080259
83. Haley NJ, Seelig DM, Zabel MD, Telling GC, Hoover EA (2009) Detection of CWD prions in urine and saliva of deer by transgenic mouse bioassay. PLoS One 4(3), e4848. doi:10.1371/journal.pone.0004848
84. Houston F, Foster JD, Chong A, Hunter N, Bostock CJ (2000) Transmission of BSE by blood transfusion in sheep. Lancet 356(9234):999–1000, doi:S0140673600027197 [pii]
85. Kariv-Inbal Z, Ben-Hur T, Grigoriadis NC, Engelstein R, Gabizon R (2006) Urine from scrapie-infected hamsters comprises low levels of prion infectivity. Neurodegener Dis 3(3):123–128. doi:10.1159/000094770, 94770 [pii]
86. Ligios C, Cancedda MG, Carta A, Santucciu C, Maestrale C, Demontis F, Saba M, Patta C, DeMartini JC, Aguzzi A, Sigurdson CJ (2011) Sheep with scrapie and mastitis transmit infectious prions through the milk. J Virol 85(2):1136–1139. doi:10.1128/JVI.02022-10, JVI.02022-10 [pii]
87. Wroe SJ, Pal S, Siddique D, Hyare H, Macfarlane R, Joiner S, Linehan JM, Brandner S, Wadsworth JD, Hewitt P, Collinge J (2006) Clinical presentation and pre-mortem diagnosis of variant Creutzfeldt-Jakob disease associated with blood transfusion: a case report. Lancet 368(9552):2061–2067. doi:10.1016/S0140-6736(06)69835-8, S0140-6736(06)69835-8 [pii]
88. Cervenakova L, Yakovleva O, McKenzie C, Kolchinsky S, McShane L, Drohan WN, Brown P (2003) Similar levels of infectivity in the blood of mice infected with human-derived vCJD and GSS strains of transmissible spongiform encephalopathy. Transfusion 43(12):1687–1694, doi:586 [pii]
89. Properzi F, Logozzi M, Abdel-Haq H, Federici C, Lugini L, Azzarito T, Cristofaro I, di Sevo D, Ferroni E, Cardone F, Venditti M, Colone M, Comoy E, Durand V, Fais S, Pocchiari M (2015) Detection of exosomal prions in blood by immunochemistry techniques. J Gen Virol 96(Pt 7):1969–1974. doi:10.1099/vir.0.000117, vir.0.000117 [pii]
90. Saa P, Yakovleva O, de Castro J, Vasilyeva I, De Paoli SH, Simak J, Cervenakova L (2014) First demonstration of transmissible spongiform encephalopathy-associated prion protein (PrPTSE) in extracellular vesicles from plasma of mice infected with mouse-adapted variant Creutzfeldt-Jakob disease by in vitro amplification. J Biol Chem 289(42):29247–29260. doi:10.1074/jbc.M114.589564, M114.589564 [pii]
91. Brundin P, Melki R, Kopito R (2010) Prion-like transmission of protein aggregates in neurodegenerative diseases. Nat Rev Mol Cell Biol 11(4):301–307. doi:10.1038/nrm2873, nrm2873 [pii]
92. Jucker M, Walker LC (2013) Self-propagation of pathogenic protein aggregates in neurodegenerative diseases. Nature 501(7465):45–51. doi:10.1038/nature12481, nature12481 [pii]
93. Renner M, Melki R (2014) Protein aggregation and prionopathies. Pathol Biol (Paris) 62(3):162–168. doi:10.1016/j.patbio.2014.01.003, S0369-8114(14)00033-9 [pii]
94. Bendor JT, Logan TP, Edwards RH (2013) The function of alpha-synuclein. Neuron 79(6):1044–1066. doi:10.1016/j.neu-

ron.2013.09.004, S0896-6273(13)00802-7 [pii]
95. Ling SC, Polymenidou M, Cleveland DW (2013) Converging mechanisms in ALS and FTD: disrupted RNA and protein homeostasis. Neuron 79(3):416–438. doi:10.1016/j.neuron.2013.07.033, S0896-6273(13)00657-0 [pii]
96. Mallucci G (2013) Spreading proteins in neurodegeneration: where do they take us? Brain 136(Pt 4):994–995. doi:10.1093/brain/awt072, awt072 [pii]
97. Mohamed NV, Herrou T, Plouffe V, Piperno N, Leclerc N (2013) Spreading of tau pathology in Alzheimer's disease by cell-to-cell transmission. Eur J Neurosci 37(12):1939–1948. doi:10.1111/ejn.12229
98. Nonaka T, Masuda-Suzukake M, Arai T, Hasegawa Y, Akatsu H, Obi T, Yoshida M, Murayama S, Mann DM, Akiyama H, Hasegawa M (2013) Prion-like properties of pathological TDP-43 aggregates from diseased brains. Cell Rep 4(1):124–134. doi:10.1016/j.celrep.2013.06.007, S2211-1247(13)00285-4 [pii]
99. Watts JC, Condello C, Stohr J, Oehler A, Lee J, DeArmond SJ, Lannfelt L, Ingelsson M, Giles K, Prusiner SB (2014) Serial propagation of distinct strains of Abeta prions from Alzheimer's disease patients. Proc Natl Acad Sci U S A 111(28):10323–10328. doi:10.1073/pnas.1408900111, 1408900111 [pii]
100. Watts JC, Giles K, Oehler A, Middleton L, Dexter DT, Gentleman SM, Dearmond SJ, Prusiner SB (2013) Transmission of multiple system atrophy prions to transgenic mice. Proc Natl Acad Sci U S A 110(48):19555–19560. doi:10.1073/pnas.1318268110, 1318268110 [pii]
101. Prusiner SB, Woerman AL, Mordes DA, Watts JC, Rampersaud R, Berry DB, Patel S, Oehler A, Lowe JK, Kravitz SN, Geschwind DH, Glidden DV, Halliday GM, Middleton LT, Gentleman SM, Grinberg LT, Giles K (2015) Evidence for alpha-synuclein prions causing multiple system atrophy in humans with parkinsonism. Proc Natl Acad Sci U S A 112(38):E5308–E5317. doi:10.1073/pnas.1514475112, 1514475112 [pii]
102. Basso M, Pozzi S, Tortarolo M, Fiordaliso F, Bisighini C, Pasetto L, Spaltro G, Lidonnici D, Gensano F, Battaglia E, Bendotti C, Bonetto V (2013) Mutant copper-zinc superoxide dismutase (SOD1) induces protein secretion pathway alterations and exosome release in astrocytes: implications for disease spreading and motor neuron pathology in amyotrophic lateral sclerosis. J Biol Chem 288(22):15699–15711. doi:10.1074/jbc.M112.425066, M112.425066 [pii]
103. Danzer KM, Kranich LR, Ruf WP, Cagsal-Getkin O, Winslow AR, Zhu L, Vanderburg CR, McLean PJ (2012) Exosomal cell-to-cell transmission of alpha synuclein oligomers. Mol Neurodegener 7:42. doi:10.1186/1750-1326-7-42, 1750-1326-7-42 [pii]
104. Ding X, Ma M, Teng J, Teng RK, Zhou S, Yin J, Fonkem E, Huang JH, Wu E, Wang X (2015) Exposure to ALS-FTD-CSF generates TDP-43 aggregates in glioblastoma cells through exosomes and TNTs-like structure. Oncotarget 6(27):24178–24191, doi:4680 [pii] 10.18632/oncotarget.4680
105. Emmanouilidou E, Melachroinou K, Roumeliotis T, Garbis SD, Ntzouni M, Margaritis LH, Stefanis L, Vekrellis K (2010) Cell-produced alpha-synuclein is secreted in a calcium-dependent manner by exosomes and impacts neuronal survival. J Neurosci 30(20):6838–6851, doi:30/20/6838 [pii] 10.1523/JNEUROSCI.5699-09.2010
106. Feiler MS, Strobel B, Freischmidt A, Helferich AM, Kappel J, Brewer BM, Li D, Thal DR, Walther P, Ludolph AC, Danzer KM, Weishaupt JH (2015) TDP-43 is intercellularly transmitted across axon terminals. J Cell Biol 211(4):897–911. doi:10.1083/jcb.201504057, jcb.201504057 [pii]
107. Gomes C, Keller S, Altevogt P, Costa J (2007) Evidence for secretion of Cu, Zn superoxide dismutase via exosomes from a cell model of amyotrophic lateral sclerosis. Neurosci Lett 428(1):43–46. doi:10.1016/j.neulet.2007.09.024, S0304-3940(07)01005-1 [pii]
108. Kunadt M, Eckermann K, Stuendl A, Gong J, Russo B, Strauss K, Rai S, Kugler S, Falomir Lockhart L, Schwalbe M, Krumova P, Oliveira LM, Bahr M, Mobius W, Levin J, Giese A, Kruse N, Mollenhauer B, Geiss-Friedlander R, Ludolph AC, Freischmidt A, Feiler MS, Danzer KM, Zweckstetter M, Jovin TM, Simons M, Weishaupt JH, Schneider A (2015) Extracellular vesicle sorting of alpha-Synuclein is regulated by sumoylation. Acta Neuropathol 129(5):695–713. doi:10.1007/s00401-015-1408-1
109. Rajendran L, Honsho M, Zahn TR, Keller P, Geiger KD, Verkade P, Simons K (2006) Alzheimer's disease beta-amyloid peptides are released in association with exosomes. Proc Natl Acad Sci U S A 103(30):11172–11177. doi:10.1073/pnas.0603838103, 0603838103 [pii]

110. Thery C, Amigorena S, Raposo G, Clayton A (2006) Isolation and characterization of exosomes from cell culture supernatants and biological fluids. Curr Protoc Cell Biol 3:22. doi:10.1002/0471143030.cb0322s30
111. Vilette D (2008) Cell models of prion infection. Vet Res 39(4):10. doi:10.1051/vetres:2007049, v08023 [pii]
112. Bosque PJ, Prusiner SB (2000) Cultured cell sublines highly susceptible to prion infection. J Virol 74(9):4377–4386
113. Butler DA, Scott MR, Bockman JM, Borchelt DR, Taraboulos A, Hsiao KK, Kingsbury DT, Prusiner SB (1988) Scrapie-infected murine neuroblastoma cells produce protease-resistant prion proteins. J Virol 62(5):1558–1564
114. Nishida N, Harris DA, Vilette D, Laude H, Frobert Y, Grassi J, Casanova D, Milhavet O, Lehmann S (2000) Successful transmission of three mouse-adapted scrapie strains to murine neuroblastoma cell lines overexpressing wild-type mouse prion protein. J Virol 74(1):320–325
115. Vorberg I, Raines A, Story B, Priola SA (2004) Susceptibility of common fibroblast cell lines to transmissible spongiform encephalopathy agents. J Infect Dis 189(3):431–439. doi:10.1086/381166, JID31043 [pii]
116. Arjona A, Simarro L, Islinger F, Nishida N, Manuelidis L (2004) Two Creutzfeldt-Jakob disease agents reproduce prion protein-independent identities in cell cultures. Proc Natl Acad Sci U S A 101(23):8768–8773. doi:10.1073/pnas.0400158101, 0400158101 [pii]
117. Archer F, Bachelin C, Andreoletti O, Besnard N, Perrot G, Langevin C, Le Dur A, Vilette D, Baron-Van Evercooren A, Vilotte JL, Laude H (2004) Cultured peripheral neuroglial cells are highly permissive to sheep prion infection. J Virol 78(1):482–490
118. Vilette D, Andreoletti O, Archer F, Madelaine MF, Vilotte JL, Lehmann S, Laude H (2001) Ex vivo propagation of infectious sheep scrapie agent in heterologous epithelial cells expressing ovine prion protein. Proc Natl Acad Sci U S A 98(7):4055–4059. doi:10.1073/pnas.061337998, 061337998 [pii]
119. Vella LJ, Hill AF (2008) Generation of cell lines propagating infectious prions and the isolation and characterization of cell-derived exosomes. Methods Mol Biol 459:69–82. doi:10.1007/978-1-59745-234-2_5
120. Taraboulos A, Serban D, Prusiner SB (1990) Scrapie prion proteins accumulate in the cytoplasm of persistently infected cultured cells. J Cell Biol 110(6):2117–2132
121. Sajnani G, Silva CJ, Ramos A, Pastrana MA, Onisko BC, Erickson ML, Antaki EM, Dynin I, Vazquez-Fernandez E, Sigurdson CJ, Carter JM, Requena JR (2012) PK-sensitive PrP is infectious and shares basic structural features with PK-resistant PrP. PLoS Pathog 8(3), e1002547. doi:10.1371/journal.ppat.1002547, PPATHOGENS-D-11-01621 [pii]
122. Arellano-Anaya ZE, Savistchenko J, Mathey J, Huor A, Lacroux C, Andreoletti O, Vilette D (2011) A simple, versatile and sensitive cell-based assay for prions from various species. PLoS One 6(5), e20563. doi:10.1371/journal.pone.0020563, PONE-D-11-05514 [pii]
123. Klohn PC, Stoltze L, Flechsig E, Enari M, Weissmann C (2003) A quantitative, highly sensitive cell-based infectivity assay for mouse scrapie prions. Proc Natl Acad Sci U S A 100(20):11666–11671. doi:10.1073/pnas.1834432100, 1834432100 [pii]

Chapter 12

Isolation of Platelet-Derived Extracellular Vesicles

Maria Aatonen, Sami Valkonen, Anita Böing, Yuana Yuana, Rienk Nieuwland, and Pia Siljander

Abstract

Platelets participate in several physiological functions, including hemostasis, immunity, and development. Additionally, platelets play key roles in arterial thrombosis and cancer progression. Given this plethora of functions, there is a strong interest of the role of platelet-derived (extracellular) vesicles (PDEVs) as functional mediators and biomarkers. Moreover, the majority of the blood-borne EVs are thought to originate from either platelets or directly from the platelet precursor cells, the megakaryocytes, which reside in the bone marrow. To circumvent confusion, we use the term PDEVs for both platelet-derived and/or megakaryocyte-derived EVs. PDEVs can be isolated from blood or from isolated platelets after activation. In this chapter, we describe all commonly used PDEV isolation methods from blood and prepurified platelets.

Key words Extracellular vesicle, Platelet, Microparticle, Exosome, Isolation, Size-exclusion chromatography, Immunobeads, Gradient centrifugation

1 Introduction

Blood contains high numbers of cell-derived extracellular vesicles (EVs). Most studies so far have shown that the majority of EVs in human blood originate from platelets or from their precursor cells in the bone marrow, the megakaryocytes [1]. Activated platelets will release EVs, and these EVs can be distinguished from those generated from megakaryocytes by exposure of typical activation markers, such as P-selectin (CD62P). By far, most investigators have used flow cytometry to study the presence of platelet-derived EVs (PDEVs) in blood or fractions thereof, because flow cytometry can establish the cellular origin of single EVs by using CD-specific antibodies. This method, however, has several limitations which should be carefully considered before application as well as interpretation of results [2–4].

Working with platelets or PDEVs presents two challenges. First, platelets are easily activated, leading to artifacts. This activation can occur during blood collection and/or handling, during isolation,

Maria Aatonen and Sami Valkonen are equally contributed with this chapter.

Andrew F. Hill (ed.), *Exosomes and Microvesicles: Methods and Protocols*, Methods in Molecular Biology, vol. 1545, DOI 10.1007/978-1-4939-6728-5_12, © Springer Science+Business Media LLC 2017

by contact with surfaces such as glass, by high shear or cold temperatures, etc. The risk of such artifacts can be reduced by adequate preanalytical steps such as the use of a needle with a large diameter for blood collection, low shear during blood collection, an effective anticoagulant, plastic tubes, and room temperature. Furthermore, inhibitors of platelet activation, such as prostaglandin E_1 can be added to further reduce the risk of activation. Second, platelets and similarly PDEVs easily form clumps or aggregates when pelleted by centrifugation. Aggregation of platelets is due to their activation, which results in fibrinogen binding and cross-linking of platelets. This is also the assumed cause for the clumping of the PDEVs. Because centrifugation itself may result in platelet activation and activation leads to fibrinogen binding, which is thought to be a prerequisite for release of PDEVs [5], centrifugation should be performed preferably under conditions where platelet activation is reduced or prevented. Therefore, low pH, the presence of compounds that inhibit platelet activation, the use of minimal centrifugal forces and minimizing the number of isolation steps requiring resuspension of pellets are recommendable when analysis of single platelets or PDEVs is needed.

2 Materials

2.1 Blood Collection for 50 mL of Whole Blood

1. Blood collection kit including disinfectants, tourniquet, needles with a minimum inner diameter of 0.84 mm (18 G), and 3 mL anticoagulant-containing tubes to collect blood for hematological parameter checking.
2. Acid citrate dextrose (ACD): 75 mM trisodium citrate, 39 mM citric acid, and 135 mM D(+)-glucose. Prepare solution in MqH_2O (Milli-Q water), set the pH to 4.5 and sterile filter the solution with 0.2 μm filter. Store in −20 °C.
3. Two 50 mL falcon tubes containing 5 mL at room temperature (RT) ACD as anticoagulant.

2.2 Platelet Isolation from 50 mL of Whole Blood

1. 10 mL of ACD (use at RT).
2. 100 μL of Prostaglandin E_1 (PGE_1 [Sigma-Aldrich]): Dissolve the PGE_1 in ethanol (100 μg/mL), prepare 50 μL aliquots in vials on ice and store aliquots immediately in −20 °C. Do not reuse.
3. 1000 mL of OptiPrep buffer: 10 mM 4-(2-hydroxyethyl)-1-piperazineethanesulfonic acid (Hepes) and 0.85 % NaCl, pH 7.4. Prepare the OptiPrep buffer by weighing 2,383 g Hepes and 8.5 g NaCl, set the pH 7.4, add to 1000 mL with Mq H_2O and filter with 0.2 μm filter (*see* **Note 1**), store RT.
4. 20 mL of 17, 13, and 10% v/v iodixanol (OptiPrep, Axis-Shield, Oslo, Norway) solutions: dilute in OptiPrep buffer.

Store RT covered from light and mix well before use. Use fresh solutions.

5. 10× stock solution of Tyrode's buffer: 1.37 M NaCl, 3 mM NaH_2PO_4, and 35 mM Hepes, pH 7.35. Prepare 10× stock solution by weighing 40 g NaCl, 0.206 g $NaH_2PO_4{\bullet}H_2O$, 4.17 g Hepes, set the pH at 7.35 and add to 500 mL with MqH_2O. Store at 4 °C.
6. 10× stock solution of glucose buffer: 55 mM D(+)-glucose. 1.982 g D(+)-glucose in 200 mL of MqH_2O, sterile filter with 0.2 μm filter and store at 4 °C.
7. 20 mL of working Tyrode's buffer: 137 mM NaCl, 0.3 mM NaH_2PO_4, 3.5 mM Hepes. Take 50 mL of 10× Tyrode's stock solution, 50 mL of 10× glucose stock solution and add to 500 mL with MqH_2O. Filter with 0.1 μm filter store at 4 °C.
8. Hematology analyzer or equivalent for counting platelets.

2.3 Generation of PDEVs by Platelet Activation

1. 100 μL of 200 mM $CaCl_2$, 500 mM KCl, and 200 mM $MgCl_2$ solutions: Prepare by dissolving salts in MqH_2O and store at RT.
2. Platelet agonists to stimulate vesiculation might include co-stimulation with collagen and thrombin or with lipopolysaccharide (LPS), LPS-binding protein (LBP), and CD14. Also stimulations with Ca^{2+}-ionophore, collagen, thrombin, cross-linked collagen related peptide (CRP-XL [6]), adenosine diphosphate (ADP), or TRAP-6 are possible (*see* **Note 2**).

2.4 Isolation of Vesicles from Human Plasma by Size-Exclusion Chromatography

1. 20 mL Sepharose CL-2B (GE Healthcare).
2. Telos filtration columns 15 mL (Kinesis).
3. 20 μm polyethylene frits (Kinesis).
4. 10 mL BD™ disposable syringe (Becton Dickinson).
5. PBS: 1.54 M NaCl, 12.4 mM Na_2HPO_4, and 2.05 mM NaH_2PO_4. Prepare in MqH_2O, set pH 7.4.
6. Trisodium citrate: 3.2% trisodium citrate in distilled water.
7. PBS/0.32% citrate: dilute 10 mL 3.2% trisodium citrate 1:10 with PBS, set pH 7.4. Filter through a 0.22 μm filter.

2.5 EV Isolation Using Antibody-Covered Beads

1. μMACS streptavidin kit consists of streptavidin-coated magnetic beads and equilibrium buffer for protein applications (Miltenyi Biotec).
2. μColumn, μMACS magnet separator, and a MultiStand (Miltenyi Biotec).
3. Hepes buffer: 10 mM Hepes, 137 mM NaCl, and 4 mM KCl. Prepare in MqH_2O, set pH 7.4.
4. Hepes–citrate buffer: 10.8 mM sodium citrate in Hepes buffer, pH 7.4.

5. Biotinylated anti-human CD41 monoclonal antibody (MoAb [Clone P2; Beckman Coulter).
6. Biotinylated mouse IgG_1 (Beckman Coulter).

3 Methods

3.1 Blood Collection

1. Collect blood from healthy donors (*see* **Note 3**).
2. To prevent platelet activation, use tourniquet lightly (*see* **Note 4**) or no tourniquet at all.
3. Discard the first 3 mL of the blood from experiments. Initial platelet count can be determined from this sample with a hematological analyzer.
4. Continue blood collection to 50 mL falcon tubes containing 5 mL of ACD (*see* **Note 5**).

3.2 Platelet Isolation from Whole Blood

1. Prepare 10 mL aliquots of whole blood in 15 mL falcon tubes and centrifuge tubes at $200\times g$, RT, 12 min without brake (*see* **Note 6**).
2. Transfer the platelet rich plasma (PRP) to 15 mL falcon tubes (*see* **Note 7**).
3. Add 10% v/v ACD and 100 ng/mL PGE_1 to PRP to prevent platelet activation.
4. Centrifuge PRP at $900\times g$, RT, 15 min.
5. Remove supernatant and softly resuspend the platelets to 2 mL of OptiPrep buffer with a 3 mL plastic Pasteur. Add to 6 mL with OptiPrep buffer. To reduce the risk of platelet activation add 100 ng/mL PGE_1 to tube (*see* **Note 8**).
6. Prepare iodixanol gradient into a 50 mL falcon tube by gently pipetting 15 mL of 17% iodixanol, 14 mL of 13% iodixanol, and 14 mL of 10% iodixanol (*see* **Note 9**).
7. Add 6 mL of platelet suspension on top of the gradient (*see* **Note 10**).
8. Centrifuge the gradients at $300\times g$, RT, 20 min without brake.
9. Recover the platelet fraction (*see* **Note 11**) from the tube and divide it to two 15 mL falcon tubes according to the volume. Add the volume to 10 mL with OptiPrep buffer, add 100 ng/mL PGE_1 and centrifuge at $900\times g$, RT, 15 min with brake.
10. Wash the obtained platelets by adding OptiPrep buffer and 100 ng/mL PGE_1.
11. Centrifuge the fractions at $900\times g$, RT, 15 min with brake.
12. Discard the supernatant and resuspend the platelets to 2–4 mL of sterile-filtered Tyrode's buffer.

13. Determine the platelet concentration and adjust it to 250×10^6 platelets/mL with sterile-filtered Tyrode's buffer (*see* **Note 12**).

3.3 Generation of PDEVs by Platelet Activation

1. Add $MgCl_2$, $CaCl_2$, and KCl solutions to the platelet suspension so that the final concentrations are 1 mM, 2 mM, and 3 mM, respectively.
2. For co-stimuli of platelets, add 1 U/mL thrombin and 10 μg/mL collagen, and for inflammatory activation add 100 ng/mL of LPS, LBP, and CD14 of each to reaction tube. Additional activations with 1 U/mL thrombin, 10 μg/mL collagen, 10 μM calcium ionophore, 1 μg/mL CRP-XL, 60 μM ADP, or 10 μM TRAP-6 can be done. After adding agonist(s), add platelet suspension to the tubes (*see* **Note 13**).
3. Incubate samples the required time in 37 °C (*see* **Note 14**).

3.4 Isolation of PDEV Subpopulations with Centrifugation

1. Centrifuge the samples at $5000 \times g$, RT, 5 min and $11{,}000 \times g$, RT, 1 min (*see* **Note 15**).
2. Transfer the supernatant to new tubes and centrifuge at $2500 \times g$, RT, 15 min.
3. Transfer the supernatant to new tubes and verify the absence of platelets and other cells with hematological analyzer. The platelet count should be zero.
4. To isolate the microparticle-enriched fraction of PDEVs, centrifuge the sample at $20{,}000 \times g$, 4 °C, for 40 min.
5. To isolate the exosome-enriched fraction of PDEVs, transfer the supernatant of **step 4** to ultracentrifuge tubes and centrifuge at $100{,}000 \times g$, 4 °C, 1–2 h (*see* **Note 16**). Remove the supernatant from the other side of the presumed location of the pellet.
6. Resuspend the pellet to PBS and centrifuge at $100{,}000 \times g$, 4 °C, 1–2 h. Remove the supernatant.
7. Use vesicles fresh or store them for further characterization in −80 °C.

3.5 EV Isolation from Human Plasma by Size-Exclusion Chromatography

1. Mix the sepharose CL-2B gently and pour 20 mL of sepharose CL-2B in a beaker. Let the sepharose settle down for at least 15 min.
2. Discard the supernatant fluid with a plastic pipet.
3. Add 15 mL PBS/0.32% citrate and swerve the mixture gently.
4. Let the sepharose settle down for at least 15 min.
5. Repeat **steps 2–4** twice.
6. Discard the supernatant and add 10 mL of PBS/0.32% citrate.

7. Place a Telos filtration column of 15 mL in a holder and level it vertically. Pipet 10 mL PBS/0.32% citrate into the column and mark the level of the buffer. Remove the air from the frit by introducing pressure to the column with the plunger of a BD syringe until the fluid drops out with steady pace (*see* **Note 17**).
8. Degas the sepharose for 5 min with vacuum. To degas PBS/0.32% citrate, filtration with 0.05 μm filter or ultrasonication can be also used.
9. Pipet the washed and degassed sepharose into the filtration column with a plastic pipet without air bubbles. Let the sepharose settle, but avoid running dry. Add more sepharose until the mark is reached with stacked sepharose. Wash the sepharose column three times with 4 mL PBS/0.32% citrate (*see* **Note 18**).
10. Place a 20 μm polyethylene frit on the bottom of a 10 mL BD syringe rough side up and remove the air from the frit with PBS/0.32% citrate as previously.
11. Remove the frit from the BD syringe by turning the syringe upside down and gently place the frit on top of the prepared sepharose column with tweezers. Please be aware that both sides of the frit are not identical. The smooth side must be placed on the sepharose. Keep the sepharose column wet with PBS/0.32% citrate until use (*see* **Notes 19** and **20**).
12. Before adding the sample wait until the PBS/citrate above the sepharose column has almost disappeared, but prevent running dry!
13. Load 1 mL of plasma on the almost dry sepharose column.
14. Immediately start collecting fractions of 500 μL. The vesicles will be eluated in fractions 8–9. However, we recommend that this is verified with a vesicle detection method, as there can be variations based on the operator/laboratory.
15. Option: Discard the first 3.5 mL. Then collect a fraction of 1 mL, which is the vesicle fraction.
16. When the plasma sample has almost completely entered the column, carefully add PBS/0.32% citrate until the column is completely filled.
17. Repeat the addition of PBS/0.32% citrate until all fractions are isolated. To complete protein and HDL elution, 26 fractions of 500 μL must be collected (*see* **Note 21**).

3.6 EV Isolation Using Antibody Covered Beads

1. Mix 100 μL of sample containing PDEVs with 0.2 μg/μL of biotinylated anti-human CD41 MoAb (*see* **Note 22**). As a negative control, biotinylated mouse IgG_1 is used at the same concentration as the antibody.
2. Rotate the mixture gently for 30 min at room temperature (RT) in the dark.

3. Add 200 μL streptavidin-coated magnetic beads to the mixture (*see* **Note 23**), rotate gently for 5 min at RT, in the dark.
4. Position the μMACS magnet separator on the MultiStand.
5. Place the μColumn on the μMACS magnet separator.
6. Rinse the μColumn once with 100 μL of the equilibration buffer and twice with 250 μL degassed Hepes–citrate buffer.
7. Add the mixture of sample with biotinylated antibody or IgG[1] and streptavidin-coated magnetic beads to the μColumn (*see* Fig. 1).
8. Collect the eluate and pass it again through the column to ensure all streptavidin-coated magnetic beads bind to the μMACS magnet separator.
9. Collect the eluate containing PDEVs-depleted fraction (negative selection) if needed for further analysis (*see* **Note 24** and Fig. 1).
10. Rinse the μColumn twice with 500 μL Hepes–citrate buffer.
11. Detach the μColumn from the μMACS magnet separator and add 75 μL Hepes–citrate buffer to collect the first eluate containing platelet EVs (positive selection, *see* Fig. 1).
12. Add 75 μL Hepes–citrate buffer again to the μColumn and quickly apply a syringe plunger (from a 5-mL syringe) to rinse the μColumn and increase elution efficiency of PDEVs.
13. Collect and combine the second eluate with the first eluate containing PDEVs (positive fraction) and use directly for further analysis (*see* **Notes 25** and **26**).

4 Notes

1. Since small particles are studied, filters should be chosen carefully, because debris may come from the filters upon elution. Filters from Merck have proven to filter solutions the best for our purposes.
2. Please bear in mind that although calcium ionophore is not a physiological agonist to induce PDEVs. Nevertheless, this is a valuable manner to produce vesicle-like structures [7].
3. The use of any medication that might affect platelet function should be avoided during 10 days before blood donation. Fasting is recommended, with normal non-caffein fluid intake.
4. If tourniquet is used, it should be released after insertion of the needle in order to collect the blood with free flow technique [7].
5. One-sixth of the final volume should be ACD, i.e., one part ACD plus five parts blood.

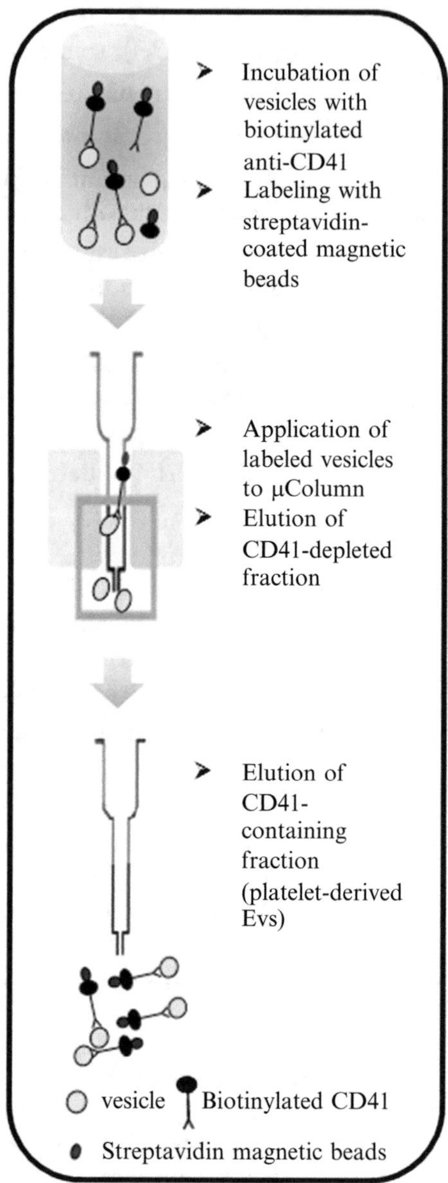

Fig. 1 General workflow for isolation of platelet-derived EVs (PDEVs) with biotinylated anti-CD41 using μMACS streptavidin beads. PDEVs-containing sample is incubated with biotinylated anti-human CD41 and labeled with streptavidin-coated magnetic beads. Sample is washed and after elution of PDEV-depleted fraction, the fraction containing CD41 positive vesicles is eluted from the column

6. When whole blood is aliquoted to tubes with smaller diameter (e.g., 15 mL falcons), usually PRP can be obtained more carefully and the buffy-coat layer is more easily avoided. Avoid blood smears at the sides of the tubes, because they may lead to poorer separation of PRP.

7. The PRP is the yellowish top phase in the tube and depending from the platelet content it can be cloudy or clearer. PRP should be obtained carefully so that no other cells are extracted among platelets. Leave ~0.5 mL of PRP on top of the cell fraction and divide the PRP in 6 mL aliquots.
8. Obtained platelets from up to 50 ml of whole blood can be put into one iodixanol gradient for separation. Therefore, the pellets are suspended to 6 mL of OptiPrep buffer.
9. The preparation of iodixanol gradient has to be done carefully. The first 15 mL of gradient (17% v/v iodixanol in OptiPrep buffer) can be pipetted directly to a 50 mL falcon tube, but the following parts of the gradient have to be layered with caution. Touch gently with the pipet tip the surface of the previous iodixanol layer. Raise the pipet tip a little, yet maintaining a contact to the previous layer through capillary action. Add the next layer of 13% iodixanol slowly and gently on top of the 17% iodixanol while raising the pipet tip constantly (while maintaining a contact to the previous layer). Repeat the same procedure with 10% iodixanol. Alternatively, you can build the gradient by letting the solutions slowly run down via the side of the slightly tilted tube. After pipetting, clear borders between the layers should be visible. If the layers are not visible, the gradient should not be used. Preparation of the gradient can be done during the centrifugation (**step 4** in Subheading 3.2) in order to place the platelet suspension to the gradient as soon as possible. After preparation, the gradient should be used within the next 20 min, otherwise the separation of the sample may not be complete.
10. The introduction of the sample to the gradient has to be done gently. Take few drops of platelet suspension to a plastic Pasteur pipette and create bubbles on top of the gradient. Pipet the rest of the platelet suspension to the gradient through the bubbles. This way the surface of the gradient is not disturbed.
11. The platelet fraction is a cloudy part in the gradient, usually about 10–14 mL in volume (*see* Fig. 2).
12. If the suspension is really cloudy, additional dilution with sterile-filtered Tyrode's buffer might be required in order to get accurate measurement.
13. Specific attention should be made to the source of for example collagen and LPS [7]. For other agonists there are several manufacturers providing products with high quality.
14. Activation times are agonist-dependent [7–11]. Note that the LPS-activation does not produce vesicles in the 10–30 min activation at 37 °C usually used for the other agonists [12].
15. Additional short spin is needed to tighten the platelet pellet if platelets are activated with agonists which do not induce strong aggregate formation (e.g. LPS) or have not been activated.

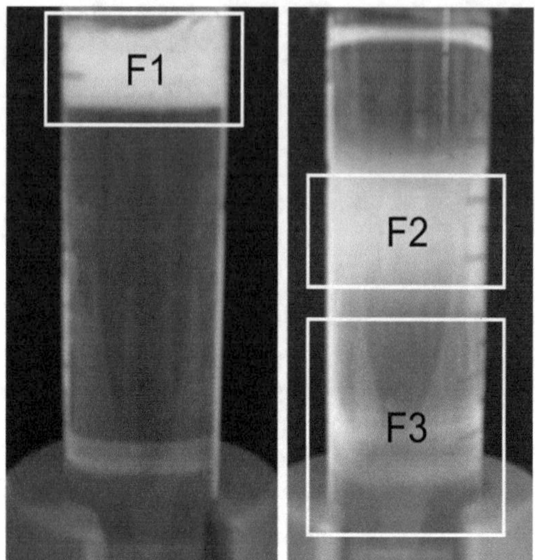

Fig. 2 The platelet fraction before and after gradient centrifugation. Fraction 1 (F1) containing platelets suspended in OptiPrep buffer before centrifugation. After centrifugation, the fraction above the platelet fraction containing the plasma/plasma protein fraction is discarded and platelets (F2) obtained. Fraction F3 showing separated leukocytes, erythrocytes, and activated platelets. Image modified from Aatonen et al. J Extracell Vesicles. 2014; 3:10.3402/jev.v3.24692

16. The centrifuge used for this purpose is Optima™ MAX-XP Ultracentrifuge (Beckman Coulter, Brea, CA, USA), combined with the rotor TLA-55, k-factor 66.
17. This procedure is necessary to degas the frit inside the filtration column.
18. The column should not run dry at any point. Make sure that at any given time there is buffer on top of the sepharose. If the column runs dry, it has to be discarded.
19. The column can be prepared on the day before experiments and stored in 4 °C, when capped and covered with Parafilm. If stored overnight, let the column reach RT and wash the column 3–4 times with buffer before loading the sample.
20. If the column is prepared more than 1 day before usage the used buffer should contain 0.05% (w/v) sodium azide as preservative.
21. Recycling of the column is not recommended.
22. The immune-magnetic beads method is optimized for direct isolation/separation of PDEVs from plasma [13]. The concentration of 27 μg biotinylated antihuman CD41 MoAb (*see* Subheading 2.5) is sufficient to label 10^7–10^{10} vesicles/mL derived from 100 μL plasma. If a sample containing higher concentration of vesicles is used, the concentration of antibody needs to be titrated.

23. An excess of streptavidin-coated magnetic beads (200 μL) is added to bind all bound and unbound biotinylated molecules in the mixture. The biotin and streptavidin bond also forms very rapidly and is stable over a wide range of pH and temperature [14]. Thus, prolonged incubation of biotinylated molecules with streptavidin-coated magnetic beads is not necessary.
24. The depletion efficiency greatly depends on the strength of the magnetic labeling. Other type of column (e.g., LD column from Miltenyi) which will retain both strongly and weakly labeled vesicles may be used to perform negative selection.
25. Using immune-magnetic beads method, recovery of the PDEVs in the positive fraction is ~80%.
26. Because the non-covalent interaction between biotin and streptavidin is the strongest interaction known with a dissociation constant, $K(\mathrm{d})$, in the order of 4×10^{-14} M [15], the magnetic beads cannot be detached from the vesicles. However, these magnetic beads are ~50 nm in diameter, and thus, their presence will not interfere with further downstream application such as functional assays [13].

Acknowledgments

Part of this work was funded by the European Metrology Research Programme (EMRP) under the Joint Research Project HLT02 (www.metves.eu). The EMRP is jointly funded by the EMRP participating countries within the European Association of National Metrology Institutes and the European Union (S.V., Y.Y., and R.N.).

Part of this work was funded by the Magnus Ehrnrooth Foundation, the Medicinska Understödsförening Liv och Hälsa r.f., the Otto Malm Foundation, and the Oscar Öflund Foundation (M.A. and P.S.). Part of the work was funded by the SalWe Research Program GET IT DONE Tekes grant nro 3986/31/2013 (S.V., P.S.).

References

1. Flaumenhaft R, Mairuhu AT, Italiano JE (2010) Platelet- and megakaryocyte-derived microparticles. Semin Thromb Hemost 36(8):881–887
2. van der Pol E, van Gemert MJ, Sturk A, Nieuwland R, van Leeuwen TG (2012) Single vs. swarm detection of microparticles and exosomes by flow cytometry. J Thromb Haemost 10(5):919–930
3. van der Pol E, Coumans FA, Grootemaat AE, Gardiner C, Sargent IL, Harrison P et al (2014) Particle size distribution of exosomes and microvesicles determined by transmission electron microscopy, flow cytometry, nanoparticle tracking analysis, and resistive pulse sensing. J Thromb Haemost 12(7):1182–1192
4. Varga Z, Yuana Y, Grootemaat AE, van der Pol E, Gollwitzer C, Krumrey M et al (2014) Towards traceable size determination of extracellular vesicles. J Extracell Vesicles 3:PMCID: PMC3916677. doi:10.3402/jev.v3.23298
5. Gemmell CH, Sefton MV, Yeo EL (1993) Platelet-derived microparticle formation involves glycoprotein IIb-IIIa. Inhibition by

RGDS and a Glanzmann's thrombasthenia defect. J Biol Chem 268(20):14586–14589

6. Siljander P, Farndale RW, Feijge MA, Comfurius P, Kos S, Bevers EM et al (2001) Platelet adhesion enhances the glycoprotein VI-dependent procoagulant response: Involvement of p38 MAP kinase and calpain. Arterioscler Thromb Vasc Biol 21(4):618–627
7. Aatonen MT, Ohman T, Nyman TA, Laitinen S, Gronholm M, Siljander PR (2014) Isolation and characterization of platelet-derived extracellular vesicles. J Extracell Vesicles 3:PMCID: PMC4125723. doi:10.3402/jev.v3.24692
8. Sims PJ, Wiedmer T, Esmon CT, Weiss HJ, Shattil SJ (1989) Assembly of the platelet prothrombinase complex is linked to vesiculation of the platelet plasma membrane. Studies in Scott syndrome: an isolated defect in platelet procoagulant activity. J Biol Chem 264(29):17049–17057
9. Ray DM, Spinelli SL, Pollock SJ, Murant TI, O'Brien JJ, Blumberg N et al (2008) Peroxisome proliferator-activated receptor gamma and retinoid X receptor transcription factors are released from activated human platelets and shed in microparticles. Thromb Haemost 99(1):86–95
10. Siljander P, Carpen O, Lassila R (1996) Platelet-derived microparticles associate with fibrin during thrombosis. Blood 87(11): 4651–4663
11. Gambim MH, do Carmo-Ade O, Marti L, Verissimo-Filho S, Lopes LR, Janiszewski M (2007) Platelet-derived exosomes induce endothelial cell apoptosis through peroxynitrite generation: experimental evidence for a novel mechanism of septic vascular dysfunction. Crit Care 11(5):107
12. Brown GT, McIntyre TM (2011) Lipopolysaccharide signaling without a nucleus: kinase cascades stimulate platelet shedding of proinflammatory IL-1beta-rich microparticles. J Immunol 186(9):5489–5496
13. Yuana Y, Osanto S, Bertina RM (2012) Use of immuno-magnetic beads for direct capture of nanosized microparticles from plasma. Blood Coagul Fibrinolysis 23(3):244–250
14. Holmberg A, Blomstergren A, Nord O, Lukacs M, Lundeberg J, Uhlen M (2005) The biotin-streptavidin interaction can be reversibly broken using water at elevated temperatures. Electrophoresis 26(3):501–510
15. Green NM (1990) Avidin and streptavidin. Methods Enzymol 184:51–67

Chapter 13

Bioinformatics Tools for Extracellular Vesicles Research

Shivakumar Keerthikumar, Lahiru Gangoda, Yong Song Gho, and Suresh Mathivanan

Abstract

Extracellular vesicles (EVs) are a class of membranous vesicles that are released by multiple cell types into the extracellular environment. This unique class of extracellular organelles which play pivotal role in intercellular communication are conserved across prokaryotes and eukaryotes. Depending upon the cell origin and the functional state, the molecular cargo including proteins, lipids, and RNA within the EVs are modulated. Owing to this, EVs are considered as a subrepertoire of the host cell and are rich reservoirs of disease biomarkers. In addition, the availability of EVs in multiple bodily fluids including blood has created significant interest in biomarker and signaling research. With the advancement in high-throughput techniques, multiple EV studies have embarked on profiling the molecular cargo. To benefit the scientific community, existing free Web-based resources including ExoCarta, EVpedia, and Vesiclepedia catalog multiple datasets. These resources aid in elucidating molecular mechanism and pathophysiology underlying different disease conditions from which EVs are isolated. Here, the existing bioinformatics tools to perform integrated analysis to identify key functional components in the EV datasets are discussed.

Key words Exosomes, Ectosomes, Bioinformatics, Extracellular vesicles, Microvesicles

1 Introduction

Extracellular vesicles (EVs) are evolutionarily conserved membranous vesicles that are released by a variety of cells into the extracellular environment [1–3]. Upon release, EVs can be taken up by recipient cells both local and distant. Whilst the exact mechanism by which the EVs communicate with the target cells are poorly understood [4], recent evidences suggest that the EVs orchestrate a multitude of cellular functions in the recipient cells. The molecular cargo within the EVs including nucleic acids, proteins and lipids regulate these cellular functions. Similar to a cell, EVs have a lipid bilayer membrane that consists of cholesterol, sphingomyelin and ceramide and are recognized to encompass lipid rafts [4, 5]. While standardization of EV nomenclature is yet to be achieved [6], based on the mode of biogenesis, EVs can be broadly classified into

Andrew F. Hill (ed.), *Exosomes and Microvesicles: Methods and Protocols*, Methods in Molecular Biology, vol. 1545,
DOI 10.1007/978-1-4939-6728-5_13, © Springer Science+Business Media LLC 2017

three main classes [7]: (a) ectosomes, (b) exosomes, and (c) apoptotic bodies.

EVs are shown to play multiple biological functions including immune response regulation [5, 8, 9], antigen presentation [10, 11], the transfer of RNA and proteins [12–14], transfer of infectious cargo [15–17], nonclassical secretion of proteins [18–20], and cell-to-cell communication [21–23]. Also, EVs have been involved in disease progression including the transfer of oncogenic proteins/RNA in cancer and pathogenic proteins between neurons involved in neurodegeneration [24, 25]. In addition, EVs are considered as reservoirs of disease biomarkers [4]. It has been previously identified that glioblastoma tumor specific exosomes containing EGFRvIII was detected in the serum of glioblastoma patients [13]. The detection of tumor-derived exosomes via a blood test has the potential to provide diagnostic value to provide a therapy response for cancer patients [13, 26].

With the recent interest on EVs coupled with the advancement of high-throughput techniques, many studies have identified the genomic, transcriptomic, proteomic, and lipidomic profiles of EVs. With the explosion of multitude of data, the bioinformatics analysis of such multidimensional data becomes critical. Here, the currently available free-to-use and commercial Web resources and stand-alone software will be discussed.

2 Online Databases for EV Research

To further facilitate the scientific community in new discoveries, three databases currently exists including ExoCarta, Vesiclepedia, and EVpedia. These databases have significantly reduced the burden on the researchers to collate the already published data manually prior to analyzing their own datasets.

2.1 ExoCarta

ExoCarta (http://www.exocarta.org) is a Web-based resource of exosomal proteins, RNA, and lipids (launched in 2009). It is manually curated by expert biologists and contains both published and unpublished exosomal studies [27]. With the involvement of the scientific community, the database will evolve into an enriched resource containing high-quality exosomal studies.

2.2 Vesiclepedia

Vesiclepedia (http://www.microvesicles.org) is a community driven compendium of proteins, RNA, and lipids in EVs. It is a continuous community annotation project with the active involvement of the EV researchers [7].

2.3 EVpedia

EVpedia (http://evpedia.info) is an integrated database of high-throughput datasets from both prokaryotic and eukaryotic vesicles [28]. EVpedia also provides an array of tools for global analysis of vesicular components.

As the purity of the population of EVs isolated depends on the isolation protocol employed, continuous research has been performed to characterize the markers that could be identified in each of the populations. The databases are routinely used for this purpose by various groups for establishing EV markers [29–36]. In all the databases, the precise isolation procedures along with the buoyant density (if provided in the respective study) allows the researchers to filter the datasets based on the isolation method or the EV type for further analysis. For example, a meta-analysis of exosomal data downloaded from ExoCarta revealed that exosomes are enriched with membrane proteins and depleted with nuclear proteins [37]. In addition, EVs are also shown to carry a host cell type specific protein/RNA signature [4, 38] and enriched with proteins implicated in oncogenesis [39] making them ideal source of disease biomarkers. EV-based protein secretion as a possible mechanism of nonclassical secretory pathway has prompted various groups to query the three databases for proteins lacking signal peptides [40, 41].

3 Software Tools for Global Analysis

As EVs carry a rich cargo of proteins, RNA, and lipids, the use of high-throughput analysis is logical (Fig. 1). To understand the precise functions of EVs, most of the studies (hypothesis and data driven) are required to perform global high-throughput analyses. Currently, many free-to-use and commercial software packages are available for analyzing the EV-based OMICS datasets (Table 1). The tools adopt a common strategy to map the provided list of genes/proteins systematically to custom databases and provide a list to biological annotations that are statistically enriched in the query dataset. The tools aid in identifying key pathways and processes regulating the biological function through enrichment analysis. Though there are several tools that can perform global analysis, some of the tools that are most commonly used will be discussed briefly.

3.1 DAVID

The Database for Annotation, Visualization and Integrated Discovery (DAVID) is one of the largely used Web-based enrichment analysis tool [42]. The users can upload the data and can perform functional annotation/gene classification. In addition, the tool also allows for cross database accession identifier conversion.

3.2 Cytoscape

Among the tools available for the analysis of genetic and protein interaction networks, Cytoscape is the most commonly used [43]. It is an open-source tool that allows for analysis and visualization of interaction networks (Fig. 2). Recent flow of multidimensional datasets and the popularity of Cytoscape among the scientific community have prompted for the development of multiple plugins,

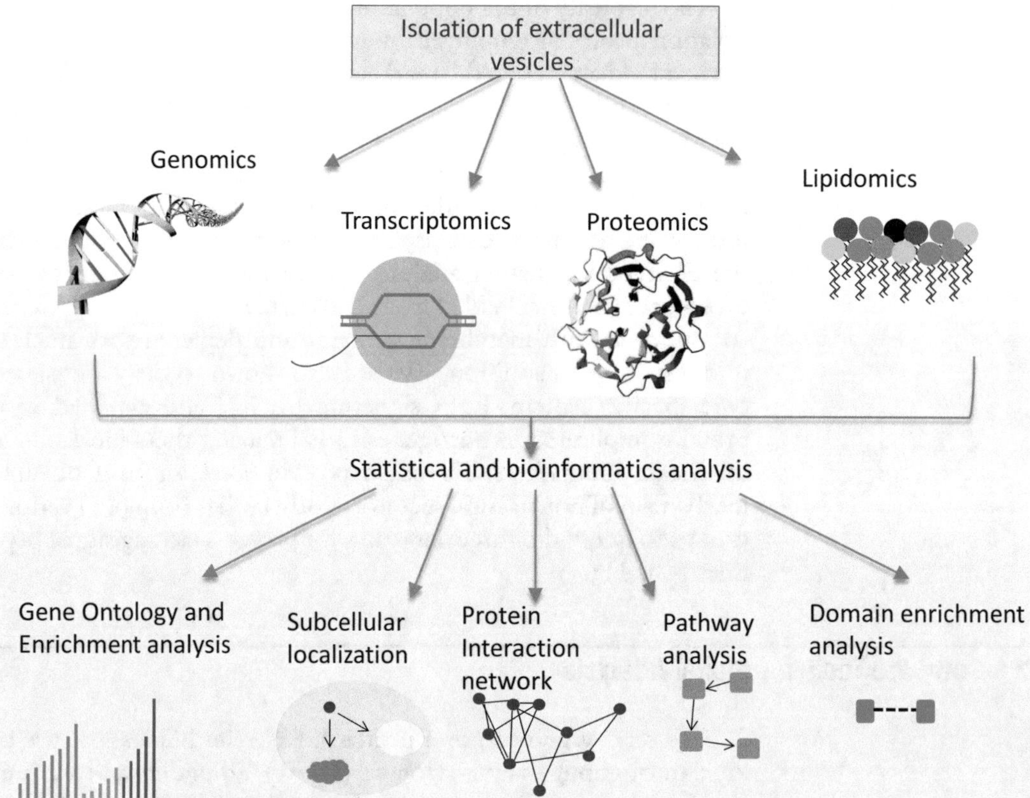

Fig. 1 A flowchart to depict the integrated bioinformatics analysis that can be performed to identify key components in EV datasets. Isolated EVs can be analyzed to profile the genomic, transcriptomic, proteomic, and lipidomic contents. With the generated datasets, multiple integrated analyses can be performed including Gene Ontology (biological process and molecular function) enrichment, protein interaction network, pathway, domain enrichment, and subcellular localization analysis

Table 1
Online tools for functional enrichment anlaysis

	Name of the tool	URL	Analysis category
1	DAVID [42]	http://david.abcc.ncifcrf.gov/	Gene ontology enrichment; pathway; protein domain
2	Cytoscape [43]	http://www.cytoscape.org/	Protein interaction network; pathway
3	IPA®	http://www.ingenuity.com/products/ipa	Gene ontology; pathway; protein interaction; disease
4	MetaCore™	http://thomsonreuters.com/metacore/	Gene ontology; pathway; protein interaction; disease

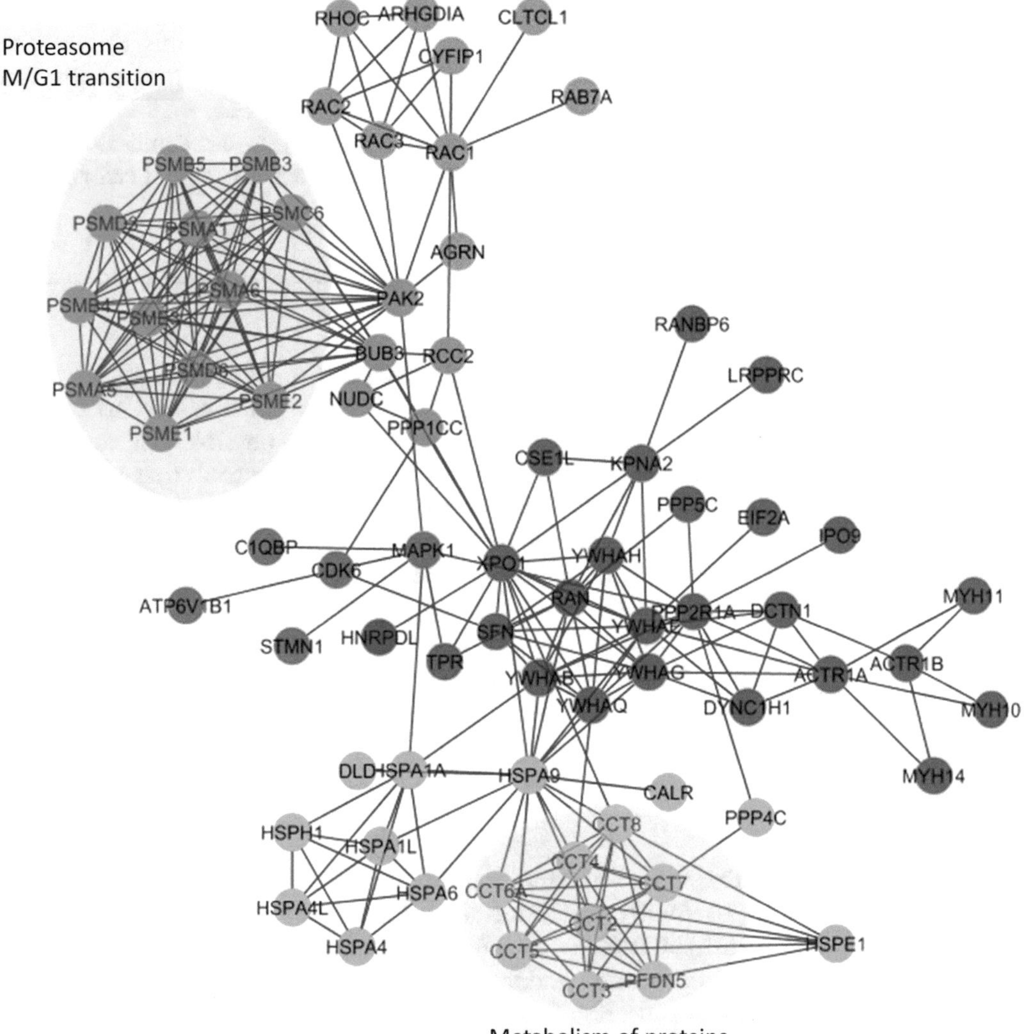

Fig. 2 A protein interaction network obtained from Cytoscape. Proteins exclusively identified in colorectal cancer cell-derived exosomes were analyzed using Cytoscape to obtain the protein interaction network. Proteins implicated in "Metabolism of proteins" and "Proteasome, M/G1 transition" were clustered in the network analysis

each through independent efforts resulting in a multifunctional tool. Users can upload the biological interaction dataset in multiple formats and visualize the interaction networks. In addition, gene ontology enrichment analysis can be performed through the plugin BiNGO [44].

3.3 Ingenuity Pathway Analysis® (IPA) and MetaCore™

IPA® and MetaCore™ are commercial software that provide the users with multiple options of analyzing large scale OMICS datasets. Both the software rely on customized datasets integrated through accumulating publicly available free-to-use scientific databases. In addition, the existing databases are enriched with additional new data through manual curation of the literature which is the major strength of these software tools. Some of the advantages and disadvantages of these two commercial softwares have been analyzed and reviewed recently [45, 46]. Nevertheless, these commercial softwares provide easy-to-use stand-alone applications allowing non-bioinformatics scientists to perform functional analysis.

4 Conclusion

With the unparalleled advancement of high-throughput experiments, bioinformatics analyses have become critical to elucidate the biological processes regulating the physiological and pathological conditions. There are many software tools that are currently available to perform global analyses. In order to understand the data better and to identify key functional modules, the users have to aware of the pitfalls associated with such tools. Most of the analyses heavily rely on the background databases that are often customized. A thorough knowledge of how the background database was assembled is important to make critical evaluation of the analysis.

Acknowledgements

This work was supported by the Australian NH&MRC fellowship (1016599) and Australian Research Council Discovery Grant (DP130100535) to S.M. The funders had no role in study design, data collection and analysis, decision to publish, or preparation of the manuscript.

References

1. Cossetti C, Smith JA, Iraci N, Leonardi T, Alfaro-Cervello C, Pluchino S (2012) Extracellular membrane vesicles and immune regulation in the brain. Front Physiol 3:117. doi:10.3389/fphys.2012.00117
2. van der Pol E, Boing AN, Harrison P, Sturk A, Nieuwland R (2012) Classification, functions, and clinical relevance of extracellular vesicles. Pharmacol Rev 64(3):676–705. doi:10.1124/pr.112.005983
3. Lee EY, Choi DY, Kim DK, Kim JW, Park JO, Kim S, Kim SH, Desiderio DM, Kim YK, Kim KP, Gho YS (2009) Gram-positive bacteria produce membrane vesicles: proteomics-based characterization of Staphylococcus aureus-derived membrane vesicles. Proteomics 9(24):5425–5436. doi:10.1002/pmic.200900338
4. Mathivanan S, Ji H, Simpson RJ (2010) Exosomes: extracellular organelles important in intercellular communication. J Proteomics 73(10):1907–1920
5. Thery C, Zitvogel L, Amigorena S (2002) Exosomes: composition, biogenesis and function. Nat Rev Immunol 2(8):569–579
6. Simpson RJ, Mathivanan S (2012) Extracellular microvesicles: the need for internationally recognised nomenclature and stringent purification criteria. J Proteomics Bioinform 5(2):ii–ii
7. Kalra H, Simpson RJ, Ji H, Aikawa E, Altevogt P, Askenase P, Bond VC, Borras FE, Breakefield X, Budnik V, Buzas E, Camussi G, Clayton A, Cocucci E, Falcon-Perez JM, Gabrielsson S, Gho YS, Gupta D, Harsha HC, Hendrix A, Hill

AF, Inal JM, Jenster G, Kramer-Albers EM, Lim SK, Llorente A, Lotvall J, Marcilla A, Mincheva-Nilsson L, Nazarenko I, Nieuwland R, Nolte-'t Hoen EN, Pandey A, Patel T, Piper MG, Pluchino S, Prasad TS, Rajendran L, Raposo G, Record M, Reid GE, Sanchez-Madrid F, Schiffelers RM, Siljander P, Stensballe A, Stoorvogel W, Taylor D, Thery C, Valadi H, van Balkom BW, Vazquez J, Vidal M, Wauben MH, Yanez-Mo M, Zoeller M, Mathivanan S (2012) Vesiclepedia: a compendium for extracellular vesicles with continuous community annotation. PLoS Biol 10(12), e1001450. doi:10.1371/journal.pbio.1001450

8. Zitvogel L, Regnault A, Lozier A, Wolfers J, Flament C, Tenza D, Ricciardi-Castagnoli P, Raposo G, Amigorena S (1998) Eradication of established murine tumors using a novel cell-free vaccine: dendritic cell-derived exosomes. Nat Med 4(5):594–600
9. Thery C, Boussac M, Veron P, Ricciardi-Castagnoli P, Raposo G, Garin J, Amigorena S (2001) Proteomic analysis of dendritic cell-derived exosomes: a secreted subcellular compartment distinct from apoptotic vesicles. J Immunol 166(12):7309–7318
10. Raposo G, Nijman HW, Stoorvogel W, Liejendekker R, Harding CV, Melief CJ, Geuze HJ (1996) B lymphocytes secrete antigen-presenting vesicles. J Exp Med 183(3):1161–1172
11. Karlsson M, Lundin S, Dahlgren U, Kahu H, Pettersson I, Telemo E (2001) "Tolerosomes" are produced by intestinal epithelial cells. Eur J Immunol 31(10):2892–2900
12. Valadi H, Ekstrom K, Bossios A, Sjostrand M, Lee JJ, Lotvall JO (2007) Exosome-mediated transfer of mRNAs and microRNAs is a novel mechanism of genetic exchange between cells. Nat Cell Biol 9(6):654–659
13. Skog J, Wurdinger T, van Rijn S, Meijer DH, Gainche L, Sena-Esteves M, Curry WT Jr, Carter BS, Krichevsky AM, Breakefield XO (2008) Glioblastoma microvesicles transport RNA and proteins that promote tumour growth and provide diagnostic biomarkers. Nat Cell Biol 10(12):1470–1476
14. Hunter MP, Ismail N, Zhang X, Aguda BD, Lee EJ, Yu L, Xiao T, Schafer J, Lee ML, Schmittgen TD, Nana-Sinkam SP, Jarjoura D, Marsh CB (2008) Detection of microRNA expression in human peripheral blood microvesicles. PLoS One 3(11), e3694
15. Robertson C, Booth SA, Beniac DR, Coulthart MB, Booth TF, McNicol A (2006) Cellular prion protein is released on exosomes from activated platelets. Blood 107(10):3907–3911
16. Fevrier B, Vilette D, Archer F, Loew D, Faigle W, Vidal M, Laude H, Raposo G (2004) Cells release prions in association with exosomes. Proc Natl Acad Sci U S A 101(26):9683–9688
17. Nguyen DG, Booth A, Gould SJ, Hildreth JE (2003) Evidence that HIV budding in primary macrophages occurs through the exosome release pathway. J Biol Chem 278(52):52347–52354
18. Amzallag N, Passer BJ, Allanic D, Segura E, Thery C, Goud B, Amson R, Telerman A (2004) TSAP6 facilitates the secretion of translationally controlled tumor protein/histamine-releasing factor via a nonclassical pathway. J Biol Chem 279(44):46104–46112
19. Chen W, Wang J, Shao C, Liu S, Yu Y, Wang Q, Cao X (2006) Efficient induction of antitumor T cell immunity by exosomes derived from heat-shocked lymphoma cells. Eur J Immunol 36(6):1598–1607
20. Zhang HG, Liu C, Su K, Yu S, Zhang L, Zhang S, Wang J, Cao X, Grizzle W, Kimberly RP (2006) A membrane form of TNF-alpha presented by exosomes delays T cell activation-induced cell death. J Immunol 176(12):7385–7393
21. Clayton A, Turkes A, Dewitt S, Steadman R, Mason MD, Hallett MB (2004) Adhesion and signaling by B cell-derived exosomes: the role of integrins. FASEB J 18(9):977–979
22. Denzer K, van Eijk M, Kleijmeer MJ, Jakobson E, de Groot C, Geuze HJ (2000) Follicular dendritic cells carry MHC class II-expressing microvesicles at their surface. J Immunol 165(3):1259–1265
23. Kim CW, Lee HM, Lee TH, Kang C, Kleinman HK, Gho YS (2002) Extracellular membrane vesicles from tumor cells promote angiogenesis via sphingomyelin. Cancer Res 62(21):6312–6317
24. Bellingham SA, Guo BB, Coleman BM, Hill AF (2012) Exosomes: vehicles for the transfer of toxic proteins associated with neurodegenerative diseases? Front Physiol 3:124. doi:10.3389/fphys.2012.00124
25. Schorey JS, Bhatnagar S (2008) Exosome function: from tumor immunology to pathogen biology. Traffic 9(6):871–881. doi:10.1111/j.1600-0854.2008.00734.x
26. Iero M, Valenti R, Huber V, Filipazzi P, Parmiani G, Fais S, Rivoltini L (2008) Tumour-released exosomes and their implications in cancer immunity. Cell Death Differ 15(1):80–88
27. Mathivanan S, Fahner CJ, Reid GE, Simpson RJ (2012) ExoCarta 2012: database of exo-

somal proteins, RNA and lipids. Nucleic Acids Res 40(Database issue):D1241–1244

28. Kim DK, Kang B, Kim OY, Choi DS, Lee J, Kim SR, Go G, Yoon YJ, Kim JH, Jang SC, Park KS, Choi EJ, Kim KP, Desiderio DM, Kim YK, Lotvall J, Hwang D, Gho YS (2013) EVpedia: an integrated database of high-throughput data for systemic analyses of extracellular vesicles. J Extracell Vesicles 2:PMID: 24009897. doi:10.3402/jev.v2i0.20384
29. Vickers KC, Palmisano BT, Shoucri BM, Shamburek RD, Remaley AT (2011) MicroRNAs are transported in plasma and delivered to recipient cells by high-density lipoproteins. Nat Cell Biol 13(4):423–433
30. Isern J, Fraser ST, He Z, Zhang H, Baron MH (2010) Dose-dependent regulation of primitive erythroid maturation and identity by the transcription factor Eklf. Blood 116(19):3972–3980
31. Koumangoye RB, Sakwe AM, Goodwin JS, Patel T, Ochieng J (2011) Detachment of breast tumor cells induces rapid secretion of exosomes which subsequently mediate cellular adhesion and spreading. PLoS One 6(9), e24234
32. Aalberts M, van Dissel-Emiliani FM, van Adrichem NP, van Wijnen M, Wauben MH, Stout TA, Stoorvogel W (2012) Identification of distinct populations of prostasomes that differentially express prostate stem cell antigen, annexin A1 and GLIPR2 in humans. Biol Reprod 86(3):82
33. Reinhardt TA, Lippolis JD, Nonnecke BJ, Sacco RE (2012) Bovine milk exosome proteome. J Proteomics 75(5):1486–1492
34. Carayon K, Chaoui K, Ronzier E, Lazar I, Bertrand-Michel J, Roques V, Balor S, Terce F, Lopez A, Salome L, Joly E (2011) Proteolipidic composition of exosomes changes during reticulocyte maturation. J Biol Chem 286(39):34426–34439
35. Lachenal G, Pernet-Gallay K, Chivet M, Hemming FJ, Belly A, Bodon G, Blot B, Haase G, Goldberg Y, Sadoul R (2011) Release of exosomes from differentiated neurons and its regulation by synaptic glutamatergic activity. Mol Cell Neurosci 46(2):409–418
36. van den Boorn JG, Picavet DI, van Swieten PF, van Veen HA, Konijnenberg D, van Veelen PA, van Capel T, Jong EC, Reits EA, Drijfhout JW, Bos JD, Melief CJ, Luiten RM (2011) Skin-depigmenting agent monobenzone induces potent T-cell autoimmunity toward pigmented cells by tyrosinase haptenation and melanosome autophagy. J Invest Dermatol 131(6):1240–1251
37. Gyorgy B, Szabo TG, Pasztoi M, Pal Z, Misjak P, Aradi B, Laszlo V, Pallinger E, Pap E, Kittel A, Nagy G, Falus A, Buzas EI (2011) Membrane vesicles, current state-of-the-art: emerging role of extracellular vesicles. Cell Mol Life Sci 68(16):2667–2688
38. Mathivanan S, Lim JW, Tauro BJ, Ji H, Moritz RL, Simpson RJ (2010) Proteomics analysis of A33 immunoaffinity-purified exosomes released from the human colon tumor cell line LIM1215 reveals a tissue-specific protein signature. Mol Cell Proteomics 9(2):197–208
39. Welton JL, Khanna S, Giles PJ, Brennan P, Brewis IA, Staffurth J, Mason MD, Clayton A (2010) Proteomics analysis of bladder cancer exosomes. Mol Cell Proteomics 9(6):1324–1338
40. Loei H, Tan HT, Lim TK, Lim KH, So JB, Yeoh KG, Chung MC (2012) Mining the gastric cancer secretome: identification of GRN as a potential diagnostic marker for early gastric cancer. J Proteome Res 11(3):1759–1772
41. Ji H, Goode RJ, Vaillant F, Mathivanan S, Kapp EA, Mathias RA, Lindeman GJ, Visvader JE, Simpson RJ (2011) Proteomic profiling of secretome and adherent plasma membranes from distinct mammary epithelial cell subpopulations. Proteomics 11(20):4029–4039
42. da Huang W, Sherman BT, Lempicki RA (2009) Systematic and integrative analysis of large gene lists using DAVID bioinformatics resources. Nat Protoc 4(1):44–57
43. Saito R, Smoot ME, Ono K, Ruscheinski J, Wang PL, Lotia S, Pico AR, Bader GD, Ideker T (2012) A travel guide to cytoscape plugins. Nat Meth 9(11):1069–1076. doi:10.1038/nmeth.2212
44. Maere S, Heymans K, Kuiper M (2005) BiNGO: a cytoscape plugin to assess overrepresentation of gene ontology categories in biological networks. Bioinformatics 21(16):3448–3449. doi:10.1093/bioinformatics/bti551
45. Muller T, Schrotter A, Loosse C, Helling S, Stephan C, Ahrens M, Uszkoreit J, Eisenacher M, Meyer HE, Marcus K (2011) Sense and nonsense of pathway analysis software in proteomics. J Proteome Res 10(12):5398–5408. doi:10.1021/pr200654k
46. Henderson-Maclennan NK, Papp JC, Talbot CC Jr, McCabe ER, Presson AP (2010) Pathway analysis software: annotation errors and solutions. Mol Genet Metab 101(2-3):134–140. doi:10.1016/j.ymgme.2010.06.005

Chapter 14

Preparation and Isolation of siRNA-Loaded Extracellular Vesicles

Pieter Vader, Imre Mäger, Yi Lee, Joel Z. Nordin, Samir E.L. Andaloussi, and Matthew J.A. Wood

Abstract

RNA interference (RNAi) has tremendous potential for specific silencing of disease-causing genes. Its clinical usage however critically depends on the development of carrier systems that can transport the RNAi-mediating small interfering RNA (siRNA) molecules to the cytosol of target cells. Recent reports have suggested that extracellular vesicles (EVs) form a natural transport system through which biomolecules, including RNA, is exchanged between cells. Therefore, EVs are increasingly being considered as potential therapeutic siRNA delivery systems.

In this chapter we describe a method for preparing siRNA-loaded EVs, including a robust, scalable method to isolate them from cell culture supernatants.

Key words Extracellular vesicles, Exosomes, Microvesicles, siRNA, Size-exclusion chromatography, Drug delivery

1 Introduction

RNA interference (RNAi) is an endogenous mechanism for regulating gene expression [1]. It is mediated by small RNA molecules, including short interfering RNA (siRNA). Because synthetic siRNA molecules introduced into cells can silence the expression of virtually any gene, their therapeutic potential is enormous. However, cellular entry for such large, hydrophilic, and anionic molecules is restricted by the plasma membrane. Therefore, before RNAi can fulfill its therapeutic potential, carrier systems that can deliver siRNA molecules to their final intracellular targets must be developed [2].

Recent reports have suggested that extracellular vesicles (EVs) form a natural transport system through which proteins, mRNA, and microRNA are exchanged between cells [3]. EVs, including exosomes and microvesicles, are membranous nanovesicles that are released from numerous cell types via multiple mechanisms [4].

Andrew F. Hill (ed.), *Exosomes and Microvesicles: Methods and Protocols*, Methods in Molecular Biology, vol. 1545, DOI 10.1007/978-1-4939-6728-5_14, © Springer Science+Business Media LLC 2017

EVs are capable of transferring RNA molecules to other cells in a selective manner, thereby influencing the phenotype and function of recipient cells [5–7]. Importantly, an increasing number of studies suggests a key role for EV-mediated intercellular communication in a variety of physiological and pathological processes [8].

Because EVs are natural carriers of small RNA, and likely utilize native mechanisms for uptake, intracellular trafficking and delivery of their content, the possibility of using EVs as vehicles for therapeutic siRNA delivery attracts increasing attention [9–11]. In fact, very encouraging proof-of-concept studies, in which EVs have been exploited as siRNA delivery vehicles, have been published recently. In these first reports, exogenous siRNA was introduced into EVs using electroporation [12, 13]. However, it was later reported that siRNA loading via electroporation is far less efficient than initially described, highlighting the need for alternative methods to prepare siRNA-loaded EVs [14]. One such alternative approach is to exploit the endogenous cellular machinery for sorting RNA cargo into EVs. This strategy involves transfection of a shRNA expression vector or synthetic siRNA into parental cells, after which siRNA-carrying EVs can be directly isolated from cell culture supernatants. Previously, these EVs have been shown to be capable of transferring functional siRNA to recipient cells [7, 15].

To date, most studies have employed differential ultracentrifugation for EV isolation [16]. However, high speed centrifugation steps cause aggregation as well as rupture of EVs, which may be detrimental for their functional recovery. We have therefore adapted and optimized an alternative EV isolation method based on ultrafiltration followed by size-exclusion chromatography, which allows high yield isolation of EVs while preserving their biophysical and functional properties.

This chapter will focus on the preparation and isolation of siRNA-loaded EVs. First, preparation of siRNA-loaded EVs using a straightforward overexpression protocol is described, followed by the methodology to isolate EVs from cell culture supernatants using size-exclusion chromatography. Furthermore, because the efficiency of siRNA loading into EVs may depend on several factors (including cell type, siRNA sequence, presence of siRNA target [17], and siRNA modifications [18]), protocols to determine loading efficiency are outlined.

2 Materials

2.1 Preparation of siRNA-Loaded Extracellular Vesicles

1. 15 cm culture dishes.
2. Dulbecco's Modified Eagle Medium (DMEM).
3. Fetal bovine serum (FBS).
4. Penicillin/streptomycin (P/S).

5. Opti-MEM (Life Technologies).
6. Lipofectamine 2000 (Life Technologies) (*see* **Note 1**).
7. siRNA-expression vector (*see* **Note 2**).
8. Phosphate-buffered saline (PBS).

2.2 Isolation of siRNA-Loaded Extracellular Vesicles

1. 0.8 μm filter membrane.
2. 100-kDa molecular weight cutoff (MWCO) Amicon spin filters (EMD Millipore).
3. Hi-Prep 16/60 Sephacryl S-400 High Resolution column (GE Healthcare) (*see* **Note 3**).
4. Chromatography system equipped with UV monitor (e.g., GE AKTA Prime Liquid Chromatography system).

2.3 Determination of siRNA Loading Efficiency

2.3.1 RNA Isolation

1. TRIZol LS reagent (Life Technologies).
2. Synthetic cel-miR-39 (cel-miR-39-3p, 5′-UCACCGGGUGU AAAUCAGCUUG-3′).
3. Chloroform.
4. Glycogen.
5. Isopropyl alcohol.
6. 75% ethanol.
7. Nuclease-free water.
8. Equipment for RNA quantification.

2.3.2 Reverse Transcription

1. Custom TaqMan Small RNA Assay, containing small RNA-specific RT and PCR primers (Life Technologies).
2. High Capacity cDNA Reverse Transcription Kit (Life Technologies).
3. Nuclease-free water.

2.3.3 PCR Amplification

1. Custom TaqMan Small RNA Assay, containing small RNA-specific RT and PCR primers (Life Technologies).
2. TaqMan Universal Master Mix II, no UNG (Life Technologies).
3. Nuclease-free water.
4. Real-time PCR instrument.

3 Methods

3.1 Preparation of siRNA-Loaded Extracellular Vesicles

1. Seed HEK293 cells (*see* **Note 4**) at a split ratio of 1:6 in 15 cm culture dishes in DMEM supplemented with 10% FBS and P/S.
2. Culture cells at 37 °C at 5% CO_2 for 24 h.

3. Remove medium, wash cells with PBS and replace with Opti-MEM.
4. Per 15 cm culture dish, dilute 40 μl Lipofectamine 2000 in 500 μl Opti-MEM. Incubate at room temperature for 5 min.
5. Per 15 cm culture dish, dilute 20 μg plasmid DNA (siRNA-expression vector) in 500 μl Opti-MEM.
6. Add Lipofectamine 2000 mixture to the plasmid DNA mixture, vortex for 10 s, and incubate at room temperature for 20 min.
7. Add Lipofectamine 2000–DNA mixture to the culture dish.
8. Incubate at 37 °C at 5 % CO_2 for 4 h.
9. Remove transfection medium, wash cells with PBS and replace with Opti-MEM supplemented with P/S (*see* **Note 5**).
10. Culture cells at 37 °C at 5 % CO_2 for 48 h, allowing cells to reach 95 % confluency.
11. Collect conditioned medium and proceed to Subheading 3.2.

3.2 Isolation of siRNA-Loaded Extracellular Vesicles

1. Centrifuge conditioned medium at 300 × *g* for 5 min, followed by 2000 × *g* for 10 min to remove cells and debris.
2. Filter conditioned medium through a 0.45 μm filter membrane (*see* **Note 6**).
3. Concentrate conditioned medium by ultrafiltration through 100-kDa molecular weight cutoff (MWCO) Amicon spin filters to a final volume of 3 ml (*see* **Note 7**).
4. Connect a Hi-Prep 16/60 Sephacryl S-400 High Resolution column to a chromatography system equipped with UV monitor (e.g., GE AKTA Prime Liquid Chromatography system) and equilibrate the column according to the manufacturer's instructions.
5. Regenerate column with one column volume of PBS at 0.5 ml/min.
6. Apply sample onto the column via a 5 ml sample loop.
7. Fractionate sample using 1.5 column volumes of PBS at 0.5 ml/min. Collect eluent into 2 ml fractions.
8. Based on UV absorbance, pool fractions containing purified EVs (*see* Fig. 1) and concentrate by ultrafiltration through 100-kDa molecular weight cutoff (MWCO) Amicon spin filters to a final volume of 200 μl.
9. Use directly or store at −80 °C for RNA isolation.

3.3 Determination of siRNA Loading Efficiency

3.3.1 RNA Isolation

1. Add 750 μl TRIzol LS reagent per 250 μl EV sample. Lyse EVs by pipetting up and down, followed by incubation at room temperature for 5 min. Proceed to **step 2** or store sample at −80 °C.
2. Add 3 μl of cel-miR-39 miRNA (1 pg/μl) (*see* **Note 8**).

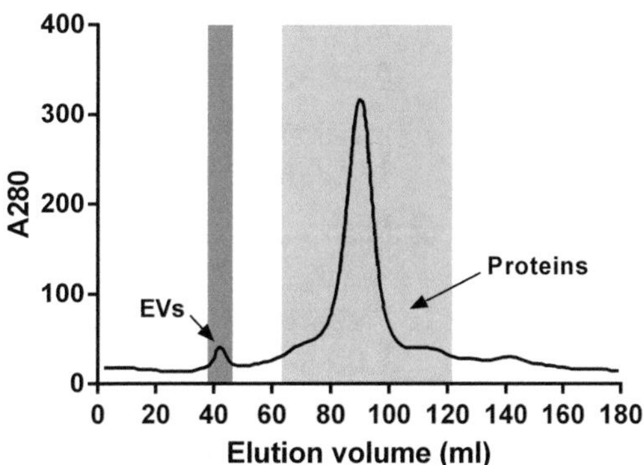

Fig. 1 Fractionation of conditioned medium. Concentrated conditioned medium was applied to a Hi-Prep 16/60 Sephacryl S-400 High Resolution column and fractionated. The eluent was monitored for UV absorbance at 280 nm. The first peak area fractions contain EVs, while later fractions contain smaller constituents, including contaminating proteins

3. Add 200 μl chloroform per 750 μl Trizol LS reagent. Shake vigorously for 15 s.
4. Incubate at room temperature for 3 min.
5. Centrifuge at 12,000 × *g* at 4 °C for 15 min.
6. Transfer the upper, aqueous phase to a new tube.
7. Add 1 μl (=20 μg) glycogen.
8. Add 500 μl isopropyl alcohol per 750 μl TRIzol LS reagent used for the homogenization in **step 1**. Vortex briefly and incubate at room temperature for 20 min.
9. Centrifuge at 12,000 × *g* at 4 °C for 10 min.
10. Discard supernatant and wash RNA pellet with 1 ml 75 % ethanol per 750 μl TRIzol LS reagent used for the homogenization in **step 1**. Vortex briefly.
11. Centrifuge at 7500 × *g* at 4 °C for 5 min.
12. Discard supernatant and air-dry the RNA pellet for 5–10 min.
13. Resuspend RNA pellet in nuclease-free water.
14. Incubate at 55–60 °C for 10–15 min.
15. Determine RNA quantity and purity (e.g., with NanoDrop and/or Quant-IT Ribogreen RNA assay).
16. Proceed to Subheading 3.3.2 or store RNA at −80 °C.

3.3.2 Reverse Transcription (RT)

1. Combine 3 μl 5× RT primer and 5 μl RNA template.
2. Incubate at 85 °C for 5 min, then at 60 °C for 5 min (*see* **Note 9**). Place on ice.
3. Prepare RT master mix on ice. Combine per reaction: 0.15 μl dNTP (100 mM), 1.00 μl RT enzyme (50 U/μl), 1.50 μl RT buffer (10×), 0.19 μl RNase inhibitor (20 U/μl) and 4.16 μl nuclease-free water.

4. Add 7 μl RT master mix to the 8 μl RT primer/denatured RNA solution from **step 2**.
5. Incubate at 16 °C for 30 min, then at 42 °C for 30 min and 85 °C for 5 min.
6. Proceed to Subheading 3.3.3 or store RT reaction at −20 °C.

3.3.3 PCR Amplification

1. Prepare PCR master mix on ice. Per 20 μl reaction, combine: 1.00 μl PCR primer (20×), 10.00 μl Taqman Universal Master Mix II (2×) and 7.67 μl nuclease-free water.
2. Combine 18.67 μl PCR master mix and 1.33 μl product from the RT reaction.
3. Incubate samples in a real-time PCR instrument at 95 °C for 10 min, followed by 40 cycles of 95 °C for 15 s and 60 °C for 1 min.
4. Normalize siRNA levels across samples using the spiked-in cel-miR-39 as controls. For absolute quantification of siRNA levels, run a standard curve using synthetic siRNA duplexes in parallel with the samples (including the RT step).

4 Notes

1. Lipofectamine 2000 can be replaced by any other alternative transfection strategy that is optimized for the cell type or cell line of interest.
2. Alternatively, cells can be transfected with synthetic siRNA duplexes. This may offer the possibility of loading chemically modified, such as fluorescent dye-labeled, siRNA duplexes, into EVs.
3. We routinely use the Hi-Prep 16/60 Sephacryl S-400 High Resolution column for isolating EVs from conditioned Opti-MEM. For purification of EVs from other sources, such as FBS-containing media or plasma, different column types and/or size exclusion media may be required.
4. In principle, any primary cell or cell line can be used as a source of siRNA-loaded EVs. We routinely use HEK293 because they are easy to grow and transfect.
5. We routinely use serum-free Opti-MEM as EV collection medium. When using cells other than HEK293 cells as source of siRNA-loaded EVs, their viability under these conditions should be verified first in order to avoid unwanted cell death. When necessary, EV-depleted FBS can be added to the culture medium (*see* **Note 3**).
6. Filtration of conditioned medium using a 0.45 μm filter membrane allows passage of all EVs smaller than 450 nm.

If exosomes are being studied specifically, the use of 0.22 μm filter membranes may be more appropriate.

7. It is recommended to wash the membrane carefully with PBS for increased EV recovery. A molecular weight cutoff of 100 kDa may be crucial for efficient recovery, because higher molecular weight cutoffs may allow passage of EVs through the membrane.
8. There is no consensus on suitable housekeeping small RNAs in EVs. Therefore, spiked-in cel-miR-39 is used to correct for differences in RNA extraction efficiency between samples.
9. When working with a double-stranded template, such as siRNA, template has to be denatured with sequence-specific RT primer before performing the reverse transcription.

Acknowledgments

P.V. was supported by a Rubicon Fellowship from the Netherlands Organisation for Scientific Research (NWO). I.M. is supported by a Postdoctoral MOBILITAS Fellowship of the Estonian Science Foundation. Y.L. is funded by the Agency of Science, Technology and Research (A*STAR), Singapore. SELA is supported by the Swedish Research Council (VR-MED and EuroNanoMedII) as well as the Swedish Society of Medical Research (SSMF).

References

1. Fire A, Xu S, Montgomery MK, Kostas SA, Driver SE, Mello CC (1998) Potent and specific genetic interference by double-stranded RNA in Caenorhabditis elegans. Nature 391(6669):806–811
2. Kanasty R, Dorkin JR, Vegas A, Anderson D (2013) Delivery materials for siRNA therapeutics. Nat Mater 12(11):967–977
3. El Andaloussi S, Mager I, Breakefield XO, Wood MJ (2013) Extracellular vesicles: biology and emerging therapeutic opportunities. Nat Rev Drug Discov 12(5):347–357
4. Raposo G, Stoorvogel W (2013) Extracellular vesicles: exosomes, microvesicles, and friends. J Cell Biol 200(4):373–383
5. Valadi H, Ekstrom K, Bossios A, Sjöstrand M, Lee JJ, Lötvall JO (2007) Exosome-mediated transfer of mRNAs and microRNAs is a novel mechanism of genetic exchange between cells. Nat Cell Biol 9(6):654–659
6. Pegtel DM, Cosmopoulos K, Thorley-Lawson DA, van Eijndhoven MA, Hopmans ES, Lindenberg JL, de Gruijl TD, Würdinger T, Middeldorp JM (2010) Functional delivery of viral miRNAs via exosomes. Proc Natl Acad Sci U S A 107(14):6328–6333
7. Kosaka N, Iguchi H, Yoshioka Y, Takeshita F, Matsuki Y, Ochiya T (2010) Secretory mechanisms and intercellular transfer of microRNAs in living cells. J Biol Chem 285(23): 17442–17452
8. van der Pol E, Boing AN, Harrison P, Sturk A, Nieuwland R (2012) Classification, functions, and clinical relevance of extracellular vesicles. Pharmacol Rev 64(3):676–705
9. Seow Y, Wood MJ (2009) Biological gene delivery vehicles: beyond viral vectors. Mol Ther 17(5):767–777
10. van Dommelen SM, Vader P, Lakhal S, Kooijmans SA, van Solinge WW, Wood MJ, Schiffelers RM (2012) Microvesicles and exosomes: opportunities for cell-derived membrane vesicles in drug delivery. J Control Release 161(2):635–644
11. Marcus ME, Leonard JN (2013) FedExosomes: engineering therapeutic biological nanoparticles

that truly deliver. Pharmaceuticals (Basel) 6(5):659–680

12. Alvarez-Erviti L, Seow Y, Yin H, Betts C, Lakhal S, Wood MJ (2011) Delivery of siRNA to the mouse brain by systemic injection of targeted exosomes. Nat Biotechnol 29(4): 341–345
13. Wahlgren J, De LKT, Brisslert M, Vaziri Sani F, Telemo E, Sunnerhagen P, Valadi H (2012) Plasma exosomes can deliver exogenous short interfering RNA to monocytes and lymphocytes. Nucleic Acids Res 40(17):e130
14. Kooijmans SA, Stremersch S, Braeckmans K, de Smedt SC, Hendrix A, Wood MJ (2013) Electroporation-induced siRNA precipitation obscures the efficiency of siRNA loading into extracellular vesicles. J Control Release 172(1):229–238
15. Zhang Y, Li L, Yu J, Zhu D, Zhang Y, Li X, Gu H, Zhang CY, Zen K (2014) Microvesicle-mediated delivery of transforming growth factor beta1 siRNA for the suppression of tumor growth in mice. Biomaterials 35(14): 4390–4400
16. Witwer KW, Buzas EI, Bemis LT, Bora A, Lässer C, Lötvall J, Nolte-'t Hoen EN, Piper MG, Sivaraman S, Skog J, Théry C, Wauben MH, Hochberg F (2013) Standardization of sample collection, isolation and analysis methods in extracellular vesicle research. J Extracell Vesicles 2
17. Squadrito ML, Baer C, Burdet F, Maderna C, Gilfillan GD, Lyle R, Ibberson M, De Palma M (2014) Endogenous RNAs modulate microRNA sorting to exosomes and transfer to acceptor cells. Cell Rep 8(5):1432–1446
18. Villarroya-Beltri C, Baixauli F, Gutierrez-Vazquez C, Sánchez-Madrid F, Mittelbrunn M (2014) Sorting it out: regulation of exosome loading. Semin Cancer Biol 28C:3–13

Chapter 15

Interaction of Extracellular Vesicles with Endothelial Cells Under Physiological Flow Conditions

Susan M. van Dommelen, Margaret Fish, Arjan D. Barendrecht, Raymond M. Schiffelers, Omolola Eniola-Adefeso, and Pieter Vader

Abstract

In the last few years it has become clear that, in addition to soluble molecules such as growth factors and cytokines, cells use extracellular vesicles (EVs) for intercellular communication. For example, EVs derived from cancer cells interact with endothelial cells, thereby affecting angiogenesis and metastasis, two essential processes in tumor progression. In most experiments, the interaction of EVs with target cells is investigated under static conditions. However the use of dynamic flow conditions is considered more relevant, especially when studying EV uptake by endothelial cells. Here, we describe the use of a perfusion system to investigate the interaction of (tumor) EVs with endothelial cells under dynamic flow conditions.

Key words Binding, Uptake, Extracellular vesicles, Physiological flow, Endothelial cells, Targeting

1 Introduction

Cells release nano-sized vesicles that are used for intercellular communication both in vitro and in vivo. These vesicles are found in bodily fluids including blood, urine, spinal fluid, breast milk, semen, and saliva. They consist of a phospholipid membrane interspersed with proteins and contain biomolecules derived from the cell of origin. Different vesicle subtypes can be discriminated based on their intracellular origin, and include exosomes and microvesicles. However, in experimental practice, it is difficult to distinguish between vesicle subgroups due to similar properties such as size and protein content. Therefore, researchers in the field have adopted the general name extracellular vesicles (EVs) [1].

Since the discovery that EVs play a role in various (patho)physiological processes, the EV research field has expanded dramatically. It is becoming clear that, in addition to soluble factors such as growth factors and cytokines, cells use EVs for intercellular communication. For example, EVs modulate immune reactions [2], affect tissue repair [3], and influence tumor progression [4].

Andrew F. Hill (ed.), *Exosomes and Microvesicles: Methods and Protocols*, Methods in Molecular Biology, vol. 1545, DOI 10.1007/978-1-4939-6728-5_15, © Springer Science+Business Media LLC 2017

The formation of new blood vessels, a process called angiogenesis, is essential to tumor growth. EVs derived from different cancer cell types were shown to activate endothelial cells, thereby driving them to a more angiogenic phenotype [5, 6]. Furthermore, tumor cell-derived EVs are able to promote metastasis by the preparation of a metastatic niche in distant organs [7, 8]. In these processes, tumor cell EVs interact with cells in the tumor environment, such as endothelial cells.

The interaction of fluorescently labeled EVs with target cells is typically studied using flow cytometry and fluorescence microscopy. In most experiments uptake and binding is studied under static conditions. However, the use of dynamic flow conditions is considered more relevant, especially when studying EV uptake by endothelial cells. Therefore, when studying the interaction of (tumor) EVs with the vessel wall, a perfusion setup is preferred over a static system.

In the protocol proposed here, fluorescently labeled tumor cell-derived EVs are perfused over endothelial cells in order to investigate their binding and uptake. We describe methods for isolation of EVs from culture supernatant, EV labeling, preparation of an endothelial cell layer, and the experimental setup of the perfusion experiment.

These methods provide opportunities to investigate the binding behavior of EVs to the vessel wall with different shear rates and in different disease or activation states of the endothelium.

2 Materials

2.1 Isolation of EVs from Cell Culture Supernatant

1. 175 cm^2 culture flasks.
2. Human epidermoid carcinoma cell line A431 (ATCC).
3. DMEM (Dulbecco's Modified Eagle Medium) high glucose cell culture medium, supplemented with 10% FBS (fetal bovine serum), 100 units/ml penicillin, and 0.1 mg/ml streptomycin.
4. EV-free FBS (*see* **Note 1**).
5. Tabletop centrifuge with temperature control.
6. 0.22 μm bottle top filter (Millipore).
7. Ultracentrifuge with appropriate rotor and tubes.
8. Phosphate buffered saline without $CaCl_2$ and $MgCl_2$ (PBS −/−), pH 7.1–7.5.

2.2 Fluorescent Labeling of EVs

1. 1 M $NaHCO_3$ solution, pH 8.3.
2. Vortex mixer.
3. Alexa Fluor 488 carboxylic acid, succinimidyl ester (Life Technologies), 10 mg/ml in DMSO.
4. Heating block for 1 ml tubes.

2.3 Preparation of Size-Exclusion Column

1. Sepharose CL-4B.
2. Column 12 cm in length, 1.6 cm in width (XK 16/20 column, GE Healthcare).
3. 20% ethanol.
4. HEPES buffered saline (HBS), 25 mM HEPES, 140 mM NaCl, pH 7.4.
5. Liquid chromatography system (e.g., ÅKTA-FPLC, GE Healthcare).

2.4 Purification of EVs After Labeling

1. 1 ml syringes.
2. 100-kDa centrifugal sample concentrator (e.g., Vivaspin, Sartorius).
3. 0.5 M NaOH solution.

2.5 Preparation of EV-Containing Perfusion Buffer

1. EV quantification system (e.g., NTA (Nanoparticle Tracking Analysis, NanoSight) or TRPS (Tunable Resistive Pulse Sensing, IZON)).
2. Phosphate buffered saline with 0.9 mM $CaCl_2$ and 0.5 mM $MgCl_2$ (PBS +/+), pH 7.1–7.5.

2.6 Preparation of Flow Perfusion Chamber

1. Silicon sheet (0.0125 cm thickness) containing 0.2 cm × 3 cm ($W \times H \times L$) perfusion channel and vacuum channels.
2. Perspex frame containing inlet and outlet tubing holders.
3. 60 °C incubator.

2.7 Preparation of an Endothelial Cell Layer

1. Glass coverslips (24 × 50 mm).
2. 96% ethanol.
3. Slide tray plate, 4 well, non-treated (e.g., PAA).
4. 1% gelatin in water, autoclaved.
5. 0.5% glutaraldehyde in water, filtered.
6. 1 M glycine in water, filtered.
7. Human umbilical vein endothelial cells (HUVECs) (Lonza).
8. EGM-2 HUVEC medium (EBM-2 medium completed with EGM-2 kit, Lonza) supplemented with 100 units/ml penicillin and 0.1 mg/ml streptomycin.

2.8 Experimental Setup and Perfusion Experiment

1. Syringes.
2. Syringe pump (e.g., Harvard apparatus 22).
3. Inlet and outlet tubings.
4. Clamp.
5. Heating block for 5 ml tubes.
6. Fluorescent microscope (e.g., Zeiss Observer Z1).
7. Vacuum pump.

3 Methods

3.1 Isolation of EVs from Cell Culture Supernatant

EVs are isolated from cell media supernatant using differential centrifugation. Isolation is preferably performed in a sterile environment (*see* **Note 2**).

1. Culture A431 cells (*see* **Note 3**) in complete cell culture medium in 3T175 flasks at 37 °C, 5 % CO_2.
2. Replace medium with EV-free medium (*see* **Note 1**) 24–48 h before cells reach 90–95 % confluency.
3. Harvest the supernatant when cells are 90–95 % confluent.
4. Centrifuge supernatant at 300 × *g* for 10 min at 4 °C to remove cells.
5. Pour supernatant into new tubes.
6. Centrifuge supernatant at 2000 × *g* for 10 min at 4 °C to remove cell debris.
7. Pour supernatant onto a 0.22 μm filter and filter under vacuum pressure.
8. Transfer supernatant to ultracentrifuge tubes and spin at 100,000 × *g* for 70 min at 4 °C to sediment EVs.
9. Aspirate supernatant and resuspend pellet in PBS −/−.
10. Spin solution at 100,000 × *g* for 70 min at 4 °C.
11. Aspirate supernatant and resuspend EV pellet in 90 μl PBS −/−.

3.2 Fluorescent Labeling of EVs

From this step on, keep EVs protected from light.

1. Add 10 μl 1 M $NaHCO_3$ to the EV sample.
2. Vortex solution.
3. Add 1 μl of Alexa Fluor 488-NHS ester (10 mg/ml in DMSO) to the solution, vortex (*see* **Note 4**).
4. Incubate EVs with the dye for 60 min at 37 °C.

3.3 Preparation of Size-Exclusion Column

1. Use CL-4B Sepharose to pack a 12 cm long and 1.6 cm wide column, according to the manufacturer's instructions. Connect the column to a liquid chromatography system to control flow rate and pressure (*see* **Note 5**).
2. Store column in 20 % ethanol at 4 °C to prevent contamination.
3. Before use, flush column with HBS and set preferred flow rate. A typical flow rate is 2.2 ml/min.

3.4 Purification of EVs After Labeling

1. Fill a 1 ml syringe with 300 μl HBS.
2. Transfer 100 μl of the stained EVs into the same syringe.
3. Inject the total volume of 400 μl onto the column.

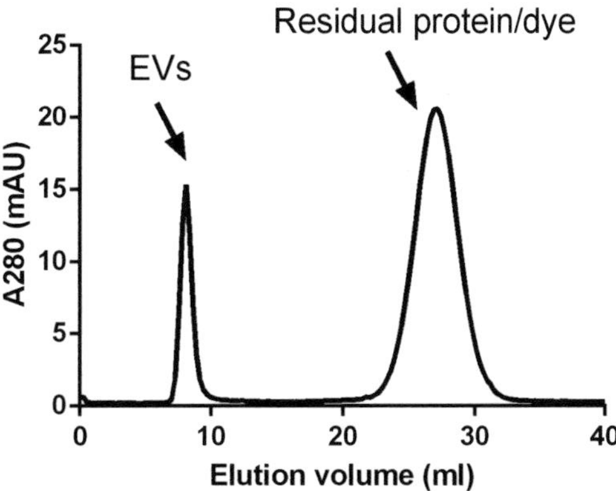

Fig. 1 UV absorbance chromatogram showing distinct fractions after EV purification using size-exclusion chromatography

4. Collect fractions containing EVs (*see* Fig. 1) and pool.
5. Change the buffer to PBS +/+ (*see* **Note 6**) using a 100-kDa centrifugal concentrator.
6. After purification, flush the column using 0.5 M NaOH.

3.5 Preparation of EV-Containing Perfusion Buffer

1. Determine the concentration of EVs in the sample using NTA or TRPS (*see* **Note 7**).
2. Dilute or concentrate (using a 100-kDa centrifugal concentrator) the sample to a final concentration of 10^{10} EVs/ml. A minimum volume of 5 ml is required.

3.6 Preparation of Flow Perfusion Chamber

The chamber consists of a perspex frame that contains the inlet and outlet of the channel formed by a silicon sheet which is placed on the frame (Fig. 2c, d).

1. One day before the experiment, pre-wet the silicon sheet with water and attach it to the perspex frame such that the channel is aligned with inlet and outlet.
2. Dry chamber overnight in a 60 °C incubator.

3.7 Preparation of an Endothelial Cell Layer

In this protocol, human umbilical vein endothelial cells (HUVECs) are used (*see* **Note 8**).

1. Sterilize glass coverslips (24×50 mm, 12.5 cm^2) using 96% ethanol.
2. Transfer each coverslip to a well in a slide tray plate and wash with PBS −/−.
3. Add 0.9 ml 1% gelatin onto each coverslip, incubate for 20 min at 37 °C, 5% CO_2.
4. Add 1.8 ml 0.5% glutaraldehyde to each coverslip, incubate at room temperature for 20–60 min.

5. Aspirate all liquid without touching the glass.
6. Add 1.8 ml 1 M glycine to each coverslip, incubate at room temperature for 20 min.
7. Aspirate all liquid without touching the glass.
8. Add 1.8 ml PBS −/− to each coverslip.
9. Aspirate all liquid without touching the glass.
10. Seed cells (0.5 ml/coverslip), incubate for 45 min at 37 °C, 5% CO_2. To obtain a confluent monolayer, seed confluent HUVECs 2 days before the experiment in a 1:2 dilution in EGM-2 (*see* **Note 9**).
11. Add 3 ml additional EGM-2 to each well.

3.8 Experimental Setup and Perfusion Experiment

The order of events in this setup (Fig. 2) is important to prevent air bubbles from entering the perfusion chamber.

1. Remove air bubbles underneath the silicon sheet by applying pressure.
2. Adjust pump settings (*see* **Note 10**).
3. Place a drop of PBS +/+ on the chamber channel to remove air from inlet and to avoid air when placing cells.
4. Fill inlet tubing with sample using a syringe, clamp tubing (*see* **Note 11**), remove syringe and connect the inlet tubing to the chamber. Keep sample in a heating block at 37 °C.
5. Connect syringe to the outlet tubing and connect outlet tubing to chamber.
6. Place coverslip on the drop of PBS +/+ with cells facing the chamber.
7. Connect vacuum and check if glass is tightly fixed to the silicon sheet.
8. Remove clamp from inlet tubing.
9. Pull sample through softly using the syringe connected to the outlet tubing.
10. Place syringe in syringe pump.
11. Start pump (*see* **Note 12**).
12. Dry outside of chamber and place it on the microscope.
13. Adjust microscope to fluorescent mode; if set correctly, a flow of fluorescent EVs should be visible.
14. Focus on the cells using bright field mode.
15. Switch to fluorescent mode.
16. Capture a video or take a time lapse of pictures. Make sure microscope stays in focus throughout the whole experiment.
17. In our experience, perfusion can be performed up to 2 h. Perfusion time is limited by loss of cell viability. Always

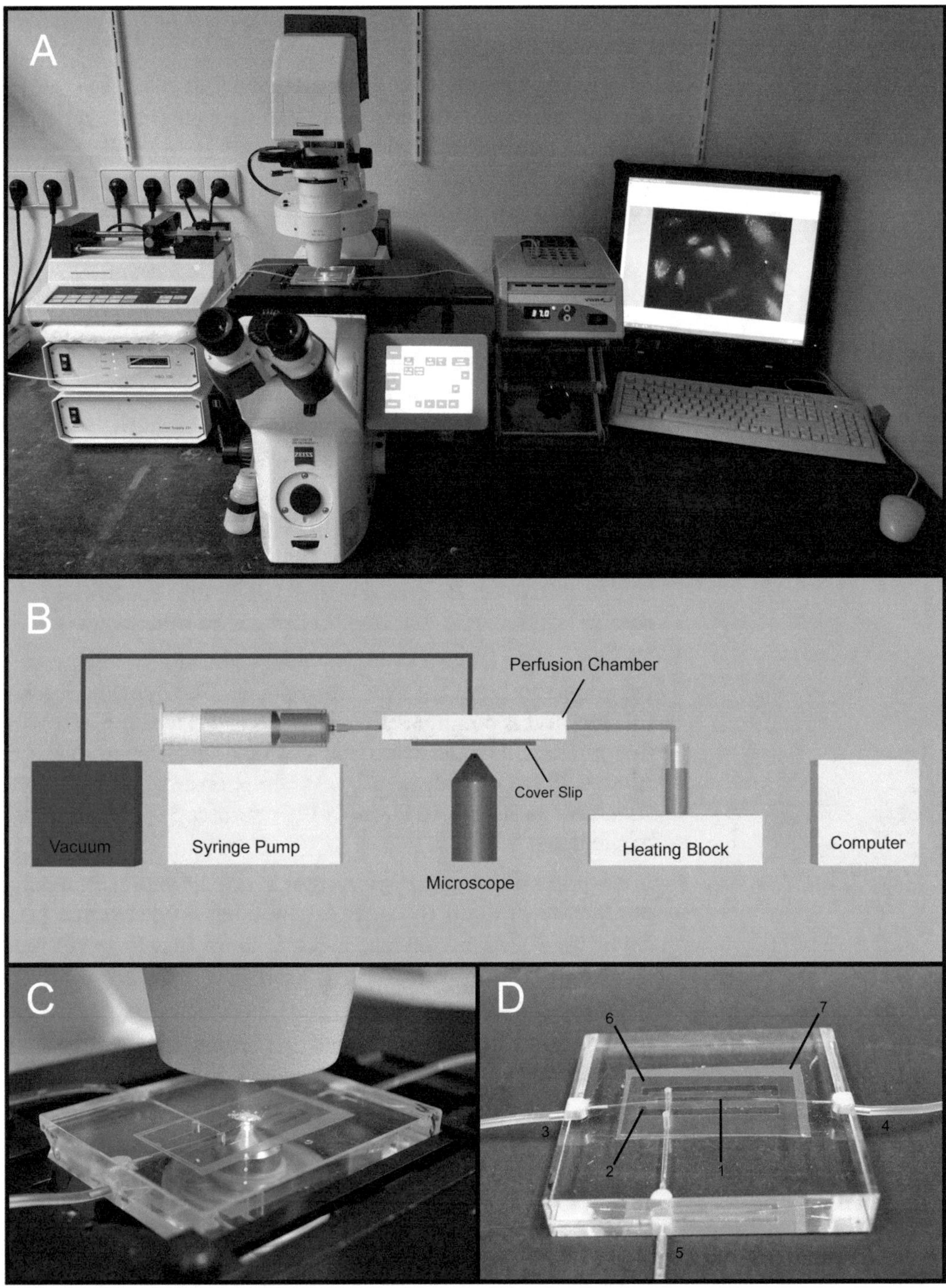

Fig. 2 Setup of the perfusion experiment. (**a**, **b**) Picture and schematic overview of experimental setup. (**c**) Picture showing the perfusion chamber placed on the microscope. (**d**) Picture showing details of the perfusion chamber. The chamber consists of a perspex frame that contains the inlet (D3) and outlet (D4) of the perfusion channel (D1), formed by a silicon sheet (D7) which is placed on the frame. The channel in the chamber we use is 0.2 cm × 0.0125 cm × 3 cm ($W \times H \times L$). A glass coverslip (D6) containing cells is placed over the channel. Vacuum (D2 vacuum channel, D5 vacuum tubing) is applied to the chamber to close the system

examine the morphology of the cells when considering experiment length.

18. After perfusion, it is possible to fixate the cells by adding fixative to the system. In that way, cells can be examined later using confocal microscopy or flow cytometry.

4 Notes

1. Fetal Bovine Serum (FBS) naturally contains EVs. Therefore, FBS is depleted from EVs by centrifuging a 30% solution of FBS in culture medium for 15–17 h at 100,000 × *g* at 4 °C. After sedimentation of the EVs, the supernatant is filtered through a 0.22 μm filter and stored at −20 °C until use. Upon use, the FBS is further diluted in culture medium.
2. For working in a sterile environment, a laminar flow cabinet should be used. Filter all the buffers before use, treat tubes and lids with 70% ethanol and allow them to dry in a flow cabinet.
3. In this protocol EVs derived from A431 human epidermoid carcinoma cells are used, but may be replaced by other types of EVs.
4. Different dyes may be used for vesicle labeling. Beside protein dyes, lipid and luminal dyes are available. The dye described in this method is conjugated to an NHS ester, which reacts with free amines. This leads to the exterior covalent coupling of Alexa 488 to proteins on EVs. When choosing a suitable dye, consider the lasers and filters of the microscope and brightness of the label.
5. If a liquid chromatography system is not available, a tabletop pump may be used to control flow rate. To determine peak fractions, UV or fluorescent measurements may be performed using a spectrofluorometer.
6. In this protocol PBS +/+ is used as perfusion buffer. Other buffers may also be considered, including culture medium, plasma, and even full blood. The magnesium and calcium ions may be crucial for binding. For example, divalent cations are crucial for ligand binding to many integrins [9].
7. NTA and TRPS provide an estimation of the number of EVs in a sample.
8. Different types of endothelial cells may be used in this method. Depending on the research question, primary cells derived from the microvasculature or aorta may be considered. Endothelial cell lines may also be used. In order to mimic a certain (disease) state, endothelial cells may be stimulated with cytokines, lipopolysaccharides or drugs before the experiment.
9. A confluent layer of endothelial cells is required for the cells to be able to resist flow.

$$\text{Shear rate} = 1.03 \times \left(\frac{6 \times \text{flow rate}}{\text{channel width} \times (\text{channel height})^2}\right)$$

Shear rate (sec^{-1})
Flow rate (ml/sec)
Channel width (cm)
Channel height (cm)

Fig. 3 Formula to convert flow rate to shear rate

10. The following settings can be adjusted in most syringe pumps: diameter of the syringe, flow direction, and flow rate. The syringe inner diameter and the pump setting determine the flow rate of the system. Therefore, calibrate the syringe pump before use. Different flow rates lead to different shear rates, depending on the size of the channel. The formula to convert flow rate to shear rate can be found in Fig. 3. A shear rate of 300 s^{-1} mimics venous and 1600 s^{-1} arterial shear rate.
11. Clamping the inlet tubing is important to prevent air from entering the chamber.
12. **Steps 7–11** need to be performed quickly to prevent static vesicle binding to the cells before perfusion starts.

Acknowledgments

The work of S.M.v.D., P.V., and R.M.S. on extracellular vesicles is supported by ERC Starting Grant 260627 'MINDS' in the FP7 Ideas program of the EU. The work of OEA on endothelial cell response to shear flow is supported by an American Heart Association Scientist Development Grant (SDG 0735043N).

References

1. Colombo M, Raposo G, Théry C (2014) Biogenesis, secretion, and intercellular interactions of exosomes and other extracellular vesicles. Annu Rev Cell Dev Biol 30:255–289
2. Robbins PD, Morelli AE (2014) Regulation of immune responses by extracellular vesicles. Nat Rev Immunol 14:195–208
3. Sluijter JPG, Verhage V, Deddens JC et al (2014) Microvesicles and exosomes for intracardiac communication. Cardiovasc Res 102:302–311
4. Vader P, Breakefield XO, Wood MJA (2014) Extracellular vesicles: emerging targets for cancer therapy. Trends Mol Med 20:385–393
5. Al-Nedawi K, Meehan B, Kerbel RS et al (2009) Endothelial expression of autocrine VEGF upon the uptake of tumor-derived microvesicles containing oncogenic EGFR. Proc Natl Acad Sci U S A 106:3794–3799
6. Skog J, Würdinger T, van Rijn S et al (2008) Glioblastoma microvesicles transport RNA and proteins that promote tumour growth and provide diagnostic biomarkers. Nat Cell Biol 10:1470–1476
7. Peinado H, Lavotshkin S, Lyden D (2011) The secreted factors responsible for pre-metastatic niche formation: old sayings and new thoughts. Semin Cancer Biol 21:139–146
8. Hood JL, San RS, Wickline SA (2011) Exosomes released by melanoma cells prepare sentinel lymph nodes for tumor metastasis. Cancer Res 71:3792–3801
9. Xiong J-P, Stehle T, Goodman SL, Arnaout MA (2003) Integrins, cations and ligands: making the connection. J Thromb Haemost 1:1642–1654

Chapter 16

Flow Cytometric Analysis of Extracellular Vesicles

Aizea Morales-Kastresana and Jennifer C. Jones

Abstract

To analyze EVs with conventional flow cytometers, most researchers will find it necessary to bind EVs to beads that are large enough to be individually resolved on the flow cytometer available in their lab or facility. Although high-resolution flow cytometers are available and are being used for EV analysis, the use of these instruments for studying EVs requires careful use and validation by experienced small-particle flow cytometrists, beyond the scope of this chapter. Shown here is a method for using streptavidin-coated beads to capture biotinylated antibodies, and stain the bead-bound EVs with directly conjugated antibodies. We find that this method is a useful tool not only on its own, without further high resolution flow cytometric analysis, but also as a means for optimizing staining methods and testing new labels for later use in high resolution, single EV flow cytometric studies. The end of the chapter includes sphere-packing calculations to quantify aspects of EV- and bead-surface geometry, as a reference for use as readers of this chapter optimize their own flow cytometry assays with EVs.

Key words Flow cytometry, Extracellular vesicles, Exosomes, Subsets

1 Introduction

High sensitivity flow cytometers have been reported [1], and methods for analysis of extracellular vesicles (EVs) have been reported with bead-based assays [2], imaging cytometers [3], and adaptations of commercially available flow cytometers [4, 5]. However, the methods of use of those instruments for the study of extracellular vesicles are specialized and not readily implemented by researchers without focused training or the assistance of experienced flow cytometrists.

The difficulty of studying EVs with flow cytometry lies in the small size of the materials being studied (Fig. 1). EVs are so much smaller than the cells that modern flow cytometers were designed to study, that the analysis of EVs with conventional flow cytometers can be accompanied by numerous artifacts if the researcher does not take appropriate precautions to avoid swarm [6], which can be due to coincident events at the laser intercept with the sample and can be due excess event anomalies in instrument signal

Andrew F. Hill (ed.), *Exosomes and Microvesicles: Methods and Protocols*, Methods in Molecular Biology, vol. 1545,
DOI 10.1007/978-1-4939-6728-5_16, © Springer Science+Business Media LLC 2017

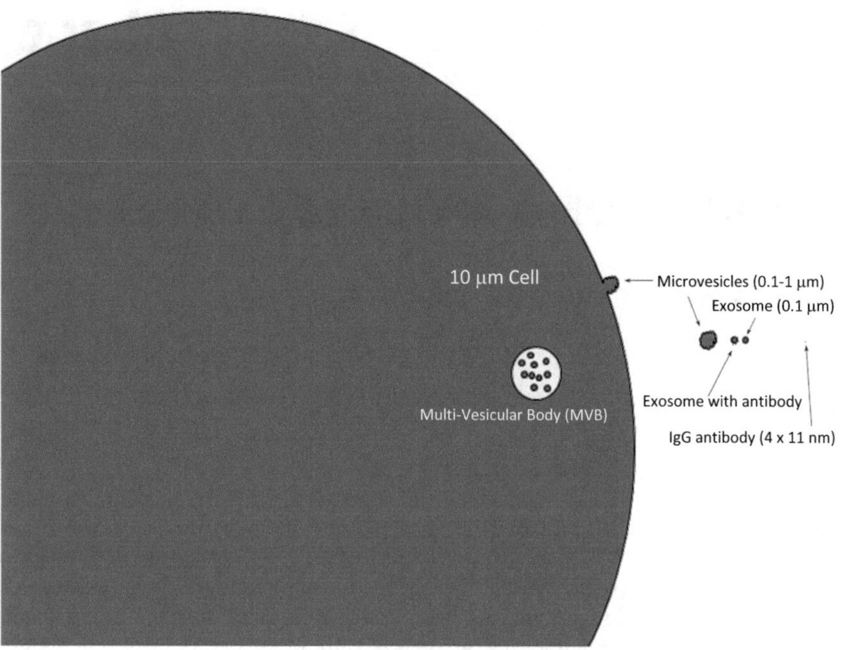

Fig. 1 Relative sizes of EVs, cells, and antibodies. The typical size of a laser intercept at the point of flow cytometric analysis of cells, beads, or EVs is 10–20 μm, while exosomes and similar EVs are ~0.1 μm. The objects in this figure are drawn to scale, to illustrate relative sizes of relevant structures and objects

processing. The purpose of this chapter is to present a method that can be used and adapted by any laboratory with access to any flow cytometer that is used to study cells.

To analyze EVs with conventional flow cytometers, a broadly useful approach is to bind EVs to beads that are large enough to be reliably resolved on the flow cytometer. An early example of this approach demonstrated that 30–100 nm exosomes could be isolated from cell culture supernatants and characterized by flow cytometry, after binding the exosomes to latex beads [7]. Shown here is a method for using streptavidin-coated beads to capture biotinylated antibodies, prior to washing the beads (to remove unbound antibodies), capturing EVs, and staining the bead-bound EVs with directly conjugated antibodies. We find that the use ligands that are specific for EV populations of interest is helpful for reducing nonspecific background that can be caused by protein binding to latex or other protein-binding beads. Figure 2 illustrates the conceptual approach, while Fig. 3 sets out the basic steps for the method. Objects in Fig. 2 are *not* drawn to scale. Rather, they are drawn to best illustrate the conceptual assembly of the beads with ligands.

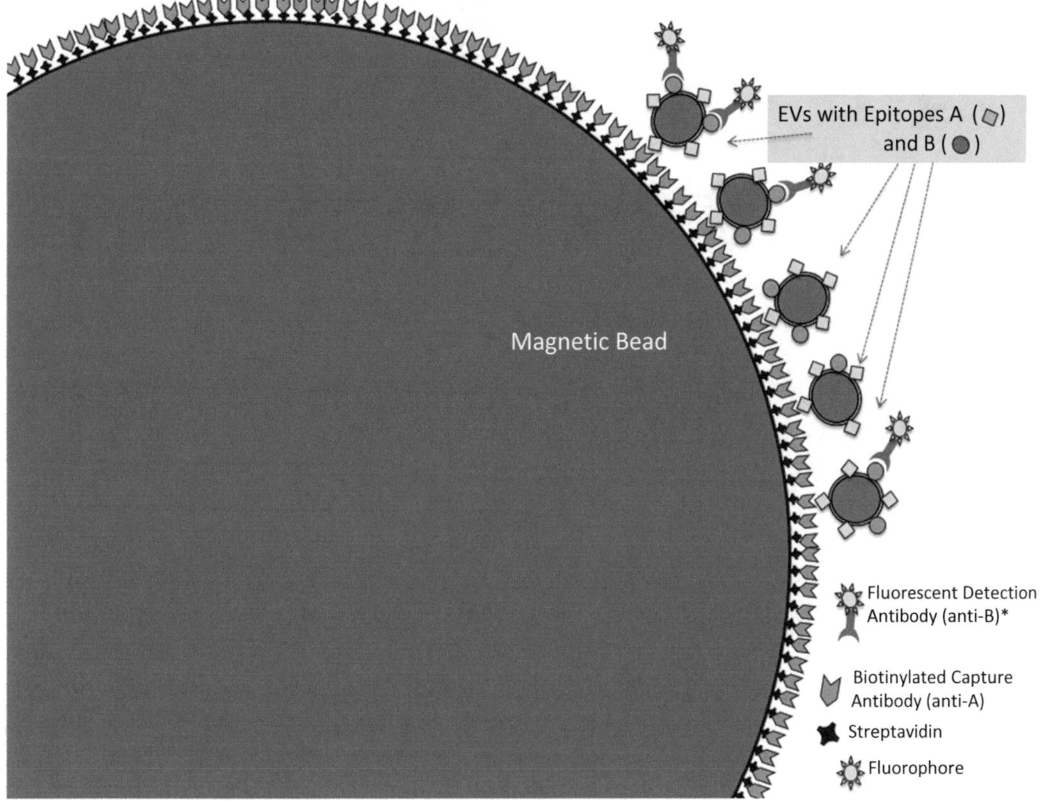

Fig. 2 Detection of EVs and EV-associated surface molecules by binding EVs to beads. To analyze EVs with conventional flow cytometers, it is generally necessary to bind EVs to beads that are large enough to be individually resolved on the flow cytometer. The objects are *not* drawn to scale. Rather, they are drawn to best illustrate the conceptual assembly of the beads with ligands

2 Materials

Maintaining sterile conditions throughout these steps will help to reduce background and preserve EV integrity. Because precipitates or small particles of salts, proteins, or other materials can interfere with nanometric sample analysis, it is important to perform all experiments with ultrapure reagents, with low background, verified by nanoparticle tracking analysis (NTA) or other similar small particle measuring instrument (*see* **Note 1**).

1. PureProteome™ Streptavidin Magnetic Beads (EMDMillipore).
2. Anti-CD81-biotin and anti-CD9-biotin, 0.5 mg/ml (Biolegend).
3. PureProteome™ Magnetic Stand, 8-well (EMDMillipore).
4. Dulbecco's hosphate buffered saline (dPBS), pH 7.4 (GIBCO/Invitrogen).

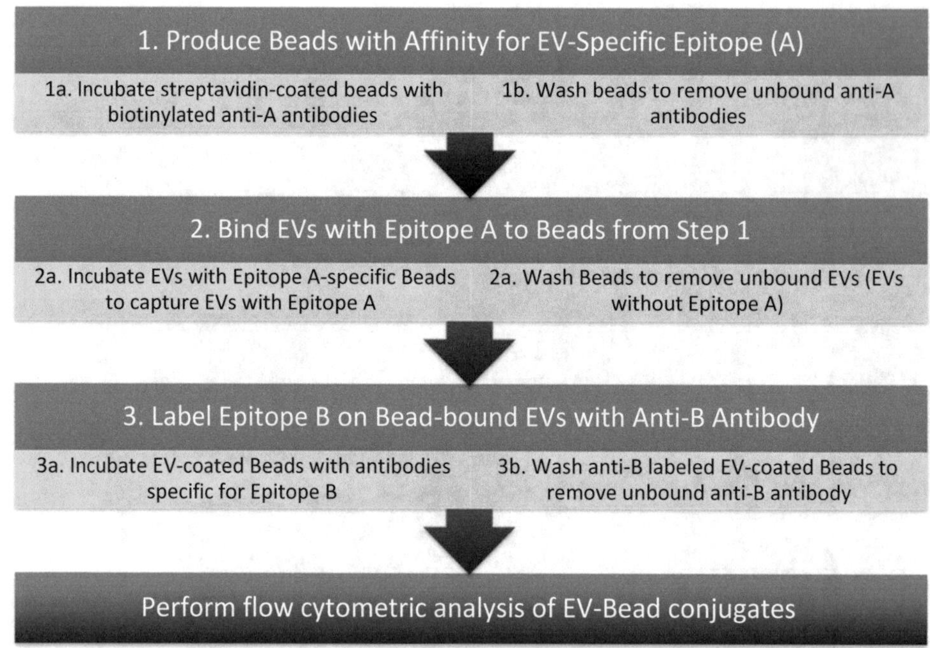

Fig. 3 Flowchart for capture and analysis of EVs by binding to beads. Shown here is a general method for using streptavidin-coated beads to capture biotinylated antibodies, prior washing the beads (to remove unbound antibodies), capturing EVs, and staining the bead-bound EVs with directly conjugated antibodies

5. Tris buffered saline (TBS), pH 7.4.
6. Tween 20 (Bio-Rad).
7. Casein blocking buffer (1% Casein in TBS).
8. T-TBS: 0.1% Tween 20 in TBS.
9. Additional antibodies for flow cytometry, including Fc-Block.

Fc-Block is Rat Anti-Mouse CD16/CD32 (BD Biosciences).

10. Agitation/mixing/inversion system.
11. DNAse/RNAse-free, sterile, low protein-binding microcentrifuge tubes.

3 Methods

There are three steps to this method. The first step is isolation and quantification of EVs, from cell culture supernatants or biofluids. Isolation of EVs can be performed in a crude manner, removing only the cells from supernatants of a cell culture, or in a more precise manner with size exclusion chromatography or with serial ultracentrifugation, as previously described [8, 9], and is not further described here. When binding EVs to beads, it is critical to know the approximate EV binding capacity of the surface of the bead, as well as the approximate EV concentration in the solution from which EVs will be captured and bound onto beads (Subheading 3.1) for staining prior to flow cytometric analysis (Subheading 3.2). Table 1 includes estimates of the relevant surface-binding capacity of cells, beads, and EVs (*see* **Notes 2** and **3**).

3.1 EV Capture on Beads

In this protocol, 10 μm magnetic beads (Millipore) are coupled with 100 μg/ml biotinylated antibody for 1 h at room temperature, under gentle agitation. EV capture is performed with rotation overnight at room temperature or at 4 °C, depending on epitope and EV-cargo stability. Beads of this size were selected due to their large EV-binding capacity and their size, which is equivalent to cells that can be visualized with standard flow cytometric methods.

This protocol is designed to prepare a minimal volume of 8 μl of beads, which we find is a minimal quantity to perform 4–5 assays, with a minimum of ~50,000 bead events to be collected for each analysis by flow cytometry (*see* **Notes 4** and **5**). Two microliters of beads can be used the capture EVs from 10 to 15 ml of tissue culture supernatants, after $300\times g$ and $10{,}000\times g$ centrifugation steps to remove cells and large debris. When adding 2 μl of the Millipore Streptavidin Magnetic Beads to 10 ml of tissue culture supernatant, the concentration of beads during the incubation is 2.4×10^4/ml. If purified (and concentrated, small volume) EV samples are used, ~10^{11} EVs at 10^{12} EV/ml is recommended as a starting point for this protocol (*see* **Note 6**).

To capture CD81-positive and CD9-positive EVs on beads:

1. Mix streptavidin-coated magnetic bead stock by inversion or gentle vortexing, to resuspend completely.
2. Transfer 2 μl of beads to a 2 ml sterile microfuge tube (round bottom, if available). Add 40 μl T-TBS, mix gently.
3. Place the tube with the beads into a magnetic stand that will support 1.5–2 ml microcentrifuge tubes (Dynal magnetic stand by Invitrogen, or PureProteome magnetic stand by EMDMillipore).
4. The beads will migrate to the magnet within minutes, and be visible as a dark (or ruddy) patch or stripe.
5. Carefully aspirate away >95 % of the liquid, being cautious not to touch the beads with the pipette tip.
6. Add 250 μl of T-TBS to the tube, and then remove the tube and mix the beads gently in the buffer to wash.
7. Place the tube back on the magnetic stand, and, again, aspirate and discard the free buffer, taking care not to disturb the beads that at the side of the tube.
8. Repeat **steps 6**–7 as a wash, twice more.
9. Add 2 μg of SAV-conjugated antibody (1 μg of anti-CD81 and 1 μg of anti-CD9; 2 μl of each antibody, in this example).

Table 1
Size vs fluorophore density for cells, EVs, and beads

Object	Typical diameter (μm)	Volume (μm^3)	Surface area (μm^2)	Max surface binding sites for 100 nm EVs[a]	Surface binding sites possible for typical IgG, short axis (5 nm)[b]	Surface binding sites possible for typical IgG, long axis (10 nm)[c]	Surface binding sites possible for PE-conjugated IgG, (~25 nm) IgG[d]
Stem cells	30–40	14,100–33,500	2800–5000	280,000 – 500,000			
Macrophages	20 - 80	4200–268,000	1250–20,000	125,000–2,000,000			
T Lymphocyte	7–10	180–520	150–314	15,000–36,000			
Microvesicles/microparticles	0.10–1	0.00005–0.52	0.03–3	–	1100–130,000	300–34,400	48–5500
Exosomes	0.04–0.12	0.00003–0.0009	0.005–0.045	–	200–2000	50–500	6–70
IgG	4 × 11 nm	–	–	–	–	–	–
10 μm bead	10	520	314	31,300–36,000	>12,550,000	>3,200,000	500,000–580,000
4 μm bead	4	33.5	50	5000–5700	>2,000,000	500,000–577,000	79,000–92,000
10 μm bead + 100 nm EVs	10.2	–	–	–	31,300 × 1100 = 34,430,000	31,300 × 300 = 9,930,000	31,300 × 48 = 1,502,400
4 μm bead + 100 nm EVs	4.2	–	–	–	5000 × 1100 = 5,500,000	5000 × 300 = 1,500,000	5000 × 48 = 240,000

Flow cytometric detection of specific epitopes on cells or EVs requires fluorescent labels, usually in the form of fluorophore-conjugated antibodies. When considering detection of fluorescently labeled EVs, as compared to the detection of fluorescently labeled cells, the relevance of size becomes apparent, especially in terms of how many antibodies can theoretically bind to the EV, cell, or bead coated with EVs. When cells or EVs are labeled with antibodies, the extremely small size of the EVs limits the number of possible epitope density. Surface binding calculations in this chart represent maximal ligand binding density, assuming complete surface area coverage with ligands with the indicated approximate ligand diameter (100 nm estimate for EVs[a]). Since antibodies are asymmetric, estimates with ligand diameters correlated with long[c] and short[b] axes of IgG were performed. Antibodies conjugated with APC (105 kDa) or PE (240 kDa) at 1:1 coupling efficiency are expected to be at least twice as large as unlabeled IgG (150 kDa)[d]

10. Add T-TBS to a final volume of 8 μl (an additional 4 μl in this example).
11. Incubate for 1 h at room temperature with gentle agitation. It is important to make sure that the solution is mixing, and that the beads are not stationary at the bottom of the tube.
12. After the incubation, place the tube back in the magnet, remove the supernatant, and repeat **steps 6** and **7** three times, with 250 μl T-TBS each time, to wash away unbound antibody. To quantify the amount of residual antibody that did not bind to the beads, the first supernatant from **step 12** may be set aside for analysis, rather than be discarded after aspiration.
13. Wash once in the magnet with phosphate buffered saline, pH 7.4.
14. Resuspend the antibody-coupled beads in 8 μl of PBS.
15. The final concentration of beads is 1.25% w/v, or approximately 120 million beads per ml.
16. Combine 2 μl of these labeled beads with 10–15 ml tissue culture supernatant that contains EVs of interest. (The supernatant should be free of cells and other debris, after a 10,000×*g* centrifugation, or equivalent size exclusion chromatography step.)
17. Incubate the beads and supernatant overnight, with constant, gentle rotation in a refrigerated room.

3.2 EV Staining on Beads for Flow Cytometric Analysis

For staining, EV-coated beads are blocked with Fc Block (Fc Block may be optional for human EVs) in a saline buffer containing 5 mg/ml casein, 25 mM Tris and 150 mM NaCl at pH 7.4, and then directly conjugated antibodies (e.g., PE-, FITC-, APC-, or other label-coupled antibodies) are added at 10 μg/ml in same buffer for 15 min (*see* **Note 7**). As with all protein-based protocols, the following protocol is best performed under refrigerated conditions (4 °C) (*see* **Notes 8** and **9**).

1. Centrifuge the mixture of supernatant with beads at 300×*g*, for 5 min. Beads and EVs bound to beads will pellet at this step.
2. Aspirate supernatant (keep for later analysis, if desired) to leave the bead pellet along with ~500 μl of fluid. The bead pellet may be difficult to visualize at this step, so leaving the 500 μl of buffer is a means of being careful not to lose the beads in the aspirate.
3. Mix the remaining 500 μl and beads, then transfer to a microfuge tube and place in the magnetic stand.
4. Remove the buffer supernatant with care to not disturb the beads at the side of the tube, beside the magnet.
5. Add, 100 μl Casein blocking buffer+2 μl Fc Block per tube.
6. Incubate for 10 min, with gentle agitation.

7. Prepare in separate tubes: 100 μl of antibody in casein-blocking buffer, for each staining assay (*see* **Note 10**).
   (a) Each staining solution should have 1 μg of antibody per tube
   (b) Example: if the concentration of the antibody stock is 100 μg/ml, use 10 μl of antibody per 100 μl of staining solution (with 90 μl casein buffer to complete the volume) (*see* **Note 11**).
   (c) For negative controls, isotype control antibodies as a "negative staining control" are one appropriate negative control, but another important negative control is a negative control for nonspecific binding to the beads. For this nonspecific binding control, we use beads that were coated with biotinylated antibody, but not EVs, and then stained with the same antibodies used to stain the EV-bound beads.
8. After 10 min blocking step with agitation, return the sample tube with beads to the magnetic stand, remove the buffer from the beads, and then add the antibody mix (*see* **Note 12**).
9. Incubate for 15 min in agitation.
10. Wash two times more with Casein-Blocking TBS buffer.
11. Resuspend the beads with stained EVs in 150 μl of blocking buffer, and proceed to analysis with conventional flow cytometric methods, with appropriate instrument calibration and sample compensation controls (Fig. 4).

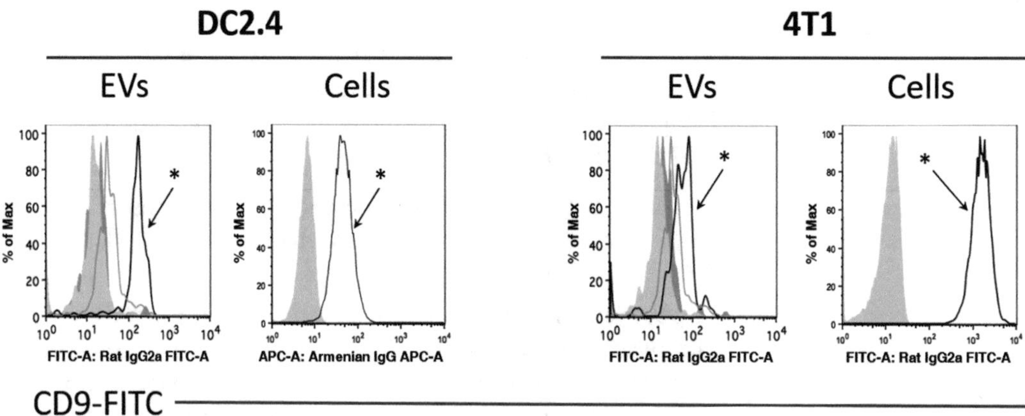

Fig. 4 Example analysis of epitope detection EVs by binding to beads. EVs isolated from DC2.4 and 4T1 cell cultures (*dark grey histograms*), as well as control EV-depleted medium (*light grey histograms*), were incubated with anti-CD9 coated magnetic beads overnight, and subsequently labeled with anti-CD9-FITC antibodies. The same clone of anti-CD9 was used for capture and detection, to ensure EV-anchored CD9 detection and not free (soluble) CD9 detection, if any. *Open histograms with* (*asterisk*) correspond to FITC-CD9 staining profile on the surface of EVs from DC2.4 and 4T1 cell lines, while the *other histograms* represent isotype and nonspecific binding controls

4 Notes

1. Antibodies, beads, EV preparations, and all combinations thereof must be titrated for optimal results. We find that saving supernatants, rather than discarding them, at steps along the protocol, and then analyzing the supernatants with protein quantification or with gel electrophoresis can help to ascertain whether more or less material may be required in future iterations.
2. We find that the Staining Index [10, 11], which is analogous to the Fisher Distance in other engineering/computational fields, is a useful statistic for comparing conditions and optimizing titrations. This statistic can be simplified as: the difference between the mean of positive population and negative (control) population, divided by the product of the standard deviation of the positive and negative populations.

$$\mathrm{SI} = \left(\mathrm{MFI}_{\mathrm{postive}} - \mathrm{MFI}_{\mathrm{negative}}\right) / \left(\mathrm{SD}_{\mathrm{positive}} \times \mathrm{SD}_{\mathrm{negative}}\right).$$

SI = Staining index.

MFI = Mean Fluorescence Intensity.

SD = Standard Deviation.

3. This bead-analysis protocol is optimized for use with cell culture supernatants that have been generated for the production of EVs. In this specific protocol, we used EVs in the range of 10^{11} EVs per bead-binding reaction. The concentration of the EVs produced by cell lines varies, depending on the cell type and on the conditions or stressors of the cell growth. As noted above, titration may be required to optimize conditions for different cell lines and for different specific EV populations that are being isolated from the supernatants.
4. Because staining intensity of the beads will depend upon number of positive EVs bound to each bead, in addition to how many epitopes are available per EV to bind to the labeled antibody, care should be taken to interpret results carefully. Brighter staining might be either due to higher levels of ligand per EV, or more EVs with the ligand bound to the bead.
5. Molecules of Equivalent Soluble Fluorescence (MESF) beads can be used to quantify number of fluorescent molecules per bead, but these beads must be run with each experiment to be quantified. MESF beads are only available (Bangs Labs or Spherotech) for certain fluorophores, such as FITC and PE, and the results can only produce estimates within the linear range of the standard curve produced by the beads.

6. IF the reader does undertake direct flow cytometric analysis of individual vesicles, additional methods may be required to remove unbound labels from the EV-bound labels. Options for this include sucrose cushions or size exclusion chromatography if dilution alone is insufficient to remove background due to the unbound label.
7. Fluorescent labels can undergo *quenching*, or diminishment of the observed fluorescence due to tight fluorophore packing. Quenching is one of the reasons that most commercial antibodies are produced to have one bound PE (phycoerythrin) molecule, rather than three or four bound PE molecules. If an antibody has too many fluorophores, the labeled antibody may appear less bright than one with an optimal coupling ratio (typically 1, or at most two PE molecules per antibody). A typical antibody is ~4 nm × 11 nm, and quenching effects that are known to be important for optimal labeling of antibody molecules should be expected to be relevant to surface labeling of 30–100 nm exosomes and other EVs as well.
8. For the methods specifically outlined here, we used dPBS, without calcium or magnesium. However, some EV epitopes, and their ligands, such as Annexin V, require calcium for binding. Selection of buffer, and inclusion or omission of cations such as calcium should be considered for this protocol, just as this would be considered for staining of cells.
9. Flow cytometer standardization with calibration beads, and appropriate compensation standards, should be performed when analyzing beads, just as with conventional flow cytometry for the analysis of cells.
10. 1 % bovine serum albumin (BSA) and 5 % BSA in PBS can be used for blocking, but we find that casein blocking buffer is more effective, and yields lower background.
11. Serum and other biofluids contain biotin, so it is preferable to link the biotinylated antibodies to the streptavidin-coated beads, prior to incubating the beads in the supernatant or biofluid, where physiological biotin would compete with the biotinylated antibody for binding to streptavidin on the beads.
12. Proteins will denature if allowed to dry. When the beads bind to the magnetic side of the microcentrifuge tube, and the supernatant is removed, buffer needs to be added within a couple of minutes to ensure that the beads do not dry out, which would denature the antibodies and proteins of the EVs, and interfere with effective staining.

References

1. Zhu S et al (2014) Light-scattering detection below the level of single fluorescent molecules for high-resolution characterization of functional nanoparticles. ACS Nano 8(10):10998–11006
2. Arakelyan A et al (2013) Nanoparticle-based flow virometry for the analysis of individual virions. J Clin Invest 123(9):3716–3727
3. Erdbrugger U, Lannigan J (2016) Analytical challenges of extracellular vesicle detection: a comparison of different techniques. Cytometry A 89(2):123–134
4. Higginbotham JN et al (2016) Identification and characterization of EGF receptor in individual exosomes by fluorescence-activated vesicle sorting. J Extracell Vesicles 5:29254
5. Danielson KM et al (2016) Diurnal variations of circulating extracellular vesicles measured by nano flow cytometry. PLoS One 11(1):e0144678
6. van der Pol E et al (2012) Single vs. swarm detection of microparticles and exosomes by flow cytometry. J Thromb Haemost 10(5):919–930
7. Lasser C, Eldh M, Lotvall J (2012) Isolation and characterization of RNA-containing exosomes. J Vis Exp 59:e3037
8. Thery C et al (2006) Isolation and characterization of exosomes from cell culture supernatants and biological fluids. Curr Protoc Cell Biol Chapter 3:22
9. Lobb RJ et al (2015) Optimized exosome isolation protocol for cell culture supernatant and human plasma. J Extracell Vesicles 4:27031
10. Baumgarth N, Bigos M (2004) Optimization of emission optics for multicolor flow cytometry. Methods Cell Biol 75:3–22
11. Maecker HT et al (2004) Selecting fluorochrome conjugates for maximum sensitivity. Cytometry A 62(2):169–173

INDEX

Andrew F. Hill (ed.), *Exosomes and Microvesicles: Methods and Protocols*, Methods in Molecular Biology, vol. 1545, DOI 10.1007/978-1-4939-6728-5, © Springer Science+Business Media LLC 2017

Q

R

S

T

U

Printed in the United States
By Bookmasters